T0361225

Chapman & Hall/CRC
Data Mining and Knowledge Discovery Series

Understanding Complex Datasets

Data Mining with Matrix Decompositions

Chapman & Hall/CRC
Data Mining and Knowledge Discovery Series

SERIES EDITOR

Vipin Kumar

University of Minnesota
Department of Computer Science and Engineering
Minneapolis, Minnesota, U.S.A

AIMS AND SCOPE

This series aims to capture new developments and applications in data mining and knowledge discovery, while summarizing the computational tools and techniques useful in data analysis. This series encourages the integration of mathematical, statistical, and computational methods and techniques through the publication of a broad range of textbooks, reference works, and handbooks. The inclusion of concrete examples and applications is highly encouraged. The scope of the series includes, but is not limited to, titles in the areas of data mining and knowledge discovery methods and applications, modeling, algorithms, theory and foundations, data and knowledge visualization, data mining systems and tools, and privacy and security issues.

PUBLISHED TITLES

UNDERSTANDING COMPLEX DATASETS: Data Mining with Matrix Decompositions
David Skillicorn

FORTHCOMING TITLES

COMPUTATIONAL METHODS OF FEATURE SELECTION
Huan Liu and Hiroshi Motoda

MULTIMEDIA DATA MINING: A Systematic Introduction to Concepts and Theory
Zhongfei Zhang and Ruofei Zhang

CONSTRAINED CLUSTERING: Advances in Algorithms, Theory, and Applications
Sugato Basu, Ian Davidson, and Kiri Wagstaff

TEXT MINING: Theory, Applications, and Visualization
Ashok Srivastava and Mehran Sahami

Chapman & Hall/CRC
Data Mining and Knowledge Discovery Series

Understanding Complex Datasets

Data Mining with Matrix Decompositions

David Skillicorn

Chapman & Hall/CRC
Taylor & Francis Group
Boca Raton London New York

Chapman & Hall/CRC is an imprint of the
Taylor & Francis Group, an **informa** business

Chapman & Hall/CRC
Taylor & Francis Group
6000 Broken Sound Parkway NW, Suite 300
Boca Raton, FL 33487-2742

© 2007 by Taylor & Francis Group, LLC
Chapman & Hall/CRC is an imprint of Taylor & Francis Group, an Informa business

No claim to original U.S. Government works

International Standard Book Number-10: 1-58488-832-6 (Hardcover)
International Standard Book Number-13: 978-1-58488-832-1 (Hardcover)

Library of Congress Cataloging-in-Publication Data

Skillicorn, David B.
   Understanding complex datasets : data mining with matrix decompositions / David Skillicorn.
      p. cm. -- (Data mining and knowledge discovery series)
   Includes bibliographical references and index.
   ISBN 978-1-58488-832-1 (alk. paper)
      1. Data mining. 2. Data structures (Computer science) 3. Computer algorithms. I. Title. II. Series.

QA76.9.D343S62 2007
005.74--dc22   2007013096

Visit the Taylor & Francis Web site at
http://www.taylorandfrancis.com

and the CRC Press Web site at
http://www.crcpress.com

For Jonathan M.D. Hill, 1968–2006

Contents

Contents

Preface

Many data-mining algorithms were developed for the world of business, fo example for customer relationship management. The datasets in this enviror ment, although large, are simple in the sense that a customer either did o did not buy three widgets, or did or did not fly from Chicago to Albuquerqu

In contrast, the datasets collected in scientific, engineering, medica and social applications often contain values that represent a combination o different properties of the real world. For example, an observation of a sta produces some value for the intensity of its radiation at a particular frequenc But the observed value is the sum of (at least) three different component the actual intensity of the radiation that the star is (was) emitting, propertie of the atmosphere that the radiation encountered on its way from the star t the telescope, and properties of the telescope itself. Astrophysicists who war to model the actual properties of stars must remove (as far as possible) th other components to get at the 'actual' data value. And it is not always clea which components are of interest. For example, we could imagine a detectio system for stealth aircraft that relied on the way they disturb the image o stellar objects behind them. In this case, a different component would be th one of interest.

Most mainstream data-mining techniques ignore the fact that real-worl datasets are *combinations* of underlying data, and build single models fror them. If such datasets can first be separated into the components that unde lie them, we might expect that the quality of the models will improve signif cantly. Matrix decompositions use the relationships among large amounts o data and the probable relationships between the components to do this kin of separation. For example, in the astrophysical example, we can plausibl assume that the changes to observed values caused by the atmosphere are ir dependent of those caused by the device. The changes in intensity might als be independent of changes caused by the atmosphere, except if the atmospher attenuates intensity non-linearly.

Some matrix decompositions have been known for over a hundred year others have only been discovered in the past decade. They are typicall computationally-intensive to compute, so it is only recently that they ha been used as analysis tools except in the most straightforward ways. Ev when matrix decompositions have been applied in sophisticated ways, th have often been used only in limited application domains, and the expe ences and 'tricks' to use them well have not been disseminated to the wic community.

This book gathers together what is known about the commonest mat decompositions:

1. Singular Value Decomposition (SVD);

2. SemiDiscrete Decomposition (SDD);

3. Independent Component Analysis (ICA);

4. Non-Negative Matrix Factorization (NNMF);

5. Tensors;

and shows how they can be used as tools to analyze large datasets. Each m trix decomposition makes a different assumption about what the underly structure in the data might be, so choosing the appropriate one is a criti choice in each application domain. Fortunately once this choice is made, m decompositions have few other parameters to set.

There are deep connections between matrix decompositions and str tures within graphs. For example, the PageRank algorithm that underlies Google search engine is related to Singular Value Decomposition, and b are related to properties of walks in graphs. Hence matrix decompositions shed light on relational data, such as the connections in the Web, or transf in the financial industry, or relationships in organizations.

This book shows how matrix decompositions can be used in practice a wide range of application domains. Data mining is becoming an import analysis tool in science and engineering in settings where controlled exp iments are impractical. We show how matrix decompositions can be u to find useful documents on the web, make recommendations about wh book or DVD to buy, look for deeply buried mineral deposits without drilli explore the structure of proteins, clean up the data from DNA microarra detect suspicious emails or cell phone calls, and figure out what topics a of documents is about.

This book is intended for researchers who have complex datasets t they want to model, and are finding that other data-mining techniques not perform well. It will also be of interest to researchers in computing w want to develop new data-mining techniques or investigate connections tween standard techniques and matrix decompositions. It can be used a supplement to graduate level data-mining textbooks.

Explanations of data mining tend to fall at two extremes. On the one hand, they reduce to "click on this button" in some data-mining software package. The problem is that a user cannot usually tell whether the algorithm that lies behind the button is appropriate for the task at hand, nor how to interpret the results that appear, or even if the results are sensible. On the other hand, other explanations require mastering a body of mathematics and related algorithms in detail. This certainly avoids the weaknesses of the software package approach, but demands a lot of the user. I have tried to steer a middle course, appropriate to a handbook. The mathematical, and to a lesser extent algorithmic, underpinnings of the data-mining techniques given here are provided, but with a strong emphasis on intuitions. My hope is that this will enable users to understand when a particular technique is appropriate and what its results mean, without having necessarily to understand every mathematical detail.

The conventional presentations of this material tend to rely on a great deal of linear algebra. Most scientists and engineers will have encountered basic linear algebra; some social scientists may have as well. For example, most will be familiar (perhaps in a hazy way) with eigenvalues and eigenvectors; but singular value decomposition is often covered only in graduate linear algebra courses, so it is not as widely known as perhaps it should be. I have tried throughout to concentrate on intuitive explanations of what the linear algebra is doing. The software that implements the decompositions described here can be used directly – there is little need to program algorithms. What is important is to understand enough about what is happening computationally to be able to set up sequences of analysis, to understand how to interpret the results, and to notice when things are going wrong.

I teach much of this material in an undergraduate data-mining course. Although most of the students do not have enough linear algebra background to understand the deeper theory behind most of the matrix decompositions, they are quickly able to learn to use them on real datasets, especially as visualization is often a natural way to interpret the results of a decomposition. I originally developed this material as background for my own graduate students who go on either to use this approach in practical settings, or to explore some of the important theoretical and algorithmic problems associated with matrix decompositions, for example reducing the computational cost.

List of Figures

Chapter 1

Data Mining

When data was primarily generated using pen and paper, there was never very much of it. The contents of the United States Library of Congress, which represent a large fraction of formal text written by humans, has been estimated to be 20TB, that is about 20 thousand billion characters. Large web search engines, at present, index about 20 billion pages, whose average size can be conservatively estimated at 10,000 characters, giving a total size of 200TB, a factor of 10 larger than the Library of Congress. Data collected about the *interactions* of people, such as transaction data and, even more so, data collected about the interactions of computers, such as message logs, can be even larger than this. Finally, there are some organizations that specialize in gathering data, for example NASA and the CIA, and these collect data at rates of about 1TB per day. Computers make it easy to collect certain kinds of data, for example transactions or satellite images, and to generate and save other kinds of data, for example driving directions. The costs of storage are so low that it is often easier to store 'everything' in case it is needed, rather than to do the work of deciding what could be deleted. The economics of personal computers, storage, and the Internet makes pack rats of us all.

The amount of data being collected and stored 'just in case' over the past two decades slowly stimulated the idea, in a number of places, that it might be useful to process such data and see what extra information might be gleaned from it. For example, the advent of computerized cash registers meant that many businesses had access to unprecedented detail about the purchasing patterns of their customers. It seemed clear that these patterns had implications for the way in which selling was done and, in particular, suggested a way of selling to each individual customer in the way that best suited him or her, a process that has come to be called *mass customization* and *customer relationship management*. Initial successes in the business com

1

text also stimulated interest in other domains where data was plentiful. example, data about highway traffic flow could be examined for ways to duce congestion; and if this worked for real highways, it could also be appl to computer networks and the Internet. Analysis of such data has beco common in many different settings over the past twenty years.

The name 'data mining' derives from the metaphor of data as someth that is large, contains far too much detail to be used as it is, but conta nuggets of useful information that can have value. So data mining can defined as the extraction of the valuable information and actionable knowle that is implicit in large amounts of data.

The data used for customer relationship management and other comm cial applications is, in a sense, quite simple. A customer either did or did purchase a particular product, make a phone call, or visit a web page. Th is no ambiguity about a value associated with a particular person, object transaction.

It is also usually true in commercial applications that a particular kin value associated to a customer or transaction, which we call an *attribute*, pl a similar role in understanding every customer. For example, the amount t a customer paid for whatever was purchased in a single trip to a store can interpreted in a similar way for every customer – we can be fairly certain t each customer wished that the amount had been smaller.

In contrast, the data collected in scientific, engineering, medical, soc and economic settings is usually more difficult to work with. The values t are recorded in the data are often a blend of several underlying proces mixed together in complex ways, and sometimes overlaid with noise. connection between a particular attribute and the structures that might l to actionable knowledge is also typically more complicated. The kinds mainstream data-mining techniques that have been successful in commer applications are less effective in these more complex settings. Matrix dec positions, the subject of this book, are a family of more-powerful techniq that can be applied to analyze complex forms of data, sometimes by the selves and sometimes as precursors to other data-mining techniques.

Much of the important scientific and technological development of last four hundred years comes from a style of investigation, probably l described by Karl Popper [91], based on controlled experiments. Research construct hypotheses inductively, but usually guided by anomalies in exist explanations of 'how things work'. Such hypotheses should have more expla tory power than existing theories, and should be easier to falsify. Suppos new hypothesis predicts that cause A is responsible for effect B. A contro experiment sets up two situations, one in which cause A is present and other in which it is not. The two situations are, as far as possible, matc with respect to all of the other variables that might influence the presenc absence of effect B. The experiment then looks at whether effect B is present only in the first situation.

Of course, few dependencies of effect on cause are perfect, so we might expect that effect B is not present in some situations where cause A is present and *vice versa*. A great deal of statistical machinery has been developed to help determine how much discrepancy can exist and still be appropriate to conclude that there is a dependency of effect B on cause A. If an experiment fails to falsify a hypothesis then this adds credibility to the hypothesis, which may eventually be promoted to a theory. Theories are not considered to be ground truth, but only approximations with useful predictiveness. This approach to understanding the universe has been enormously successful.

However, it is limited by the fact that there are four kinds of settings where controlled experiments are not directly possible:

- We do not *have access* to the variables that we would like to control. Controlled experiments are only possible on earth or its near vicinity. Understanding the wider universe cannot, at present, be achieved by controlled experiments because we cannot control the position, interactions and outputs of stars, galaxies, and other celestial objects. We can observe such objects, but we have no way to set them up in a experimental configuration.

- We do not know *how to set* the values of variables that we wish to control. Some processes are not well enough understood for us to create experimental configurations on demand. For example, fluid flowing next to a boundary will occasionally throw off turbulent eddies. However, it is not known how to make this happen. Studying the structure of such eddies requires waiting for them to happen, rather than making them happen.

- It would be *unethical* to set some variables to some values. Controlled medical experiments on human subjects can only take place if the expected differences between the control and treatment groups are small. If the treatment turns out to be either surprisingly effective or dangerously ineffective, the experiment must be halted on ethical grounds.

- The values of some variables *come from the autonomous actions* of humans. Controlled experiments in social, political, and economic settings cannot be constructed because the participants act in their own interests, regardless of the desires of the experimenters. Governments and bureaucrats have tried to avoid these limitations by trying to compel the 'right' behavior by participants, but this has been notably unsuccessful.

Controlled experiments require very precise collection of data, capturing the presence or absence of a supposed cause and the corresponding effect.

with all other variable values or attributes either held constant, or match between the two possibilities. In situations where controlled experiments not possible, such different configurations cannot be *created* to order, but th may nevertheless be present in data collected about the system of interest. example, even though we cannot make stars behave in certain ways, we may able to find two situations where the presence and absence of a hypothesi cause can be distinguished. The data from such situations can be analyzed see whether the expected relationship between cause and effect is support These are called *natural experiments*, in contrast to controlled experiment

In natural experiments, it may often be more difficult to make sure t the values of other variables or attributes are properly matched, but this be compensated for, to some extent, by the availability of a larger amount data than could be collected in a controlled experiment. More sophistica methods for arguing that dependencies imply causality are also needed.

Data mining provides techniques for this second kind of analysis, of s tems too complex or inaccessible for controlled experiments. Data mining therefore a powerful methodology for exploring systems in science, engine ing, medicine, and human society (economics, politics, social sciences, business). It is rapidly becoming an important, central tool for increasing understanding of the physical and social worlds.

1.1   What is data like?

Given a complex system, many kinds of data about it can be collected. data we will consider will usually be in the form of a set of records, each which describes one object in the system. These objects might be physi objects, for example, stars; people, for example, customers; or transactic for example, purchases at a store.

Each record contains the values for a set of *attributes* associated with record. For example, an attribute for a star might be its observed intensity a particular wavelength; an attribute for a person might be his or her heig an attribute for a transaction might be the total dollar value.

Such data can be arranged as a matrix, with one row for each obj one column for each attribute, and entries that specify the attribute val belonging to each object.

Other data formats are possible. For example, every record might have values for every attribute – a medical dataset contains information ab pregnancies only for those records corresponding to females. Such data d not trivially fit the matrix template since not every row has the same leng Another common data format is a graph, in which the connections or li between the records contain the important information. For example, a gr

of telephone calls, in which the nodes are people and the edges represent cal.
between them, can be used to detect certain kinds of fraud. Such a grap
does not trivially fit the matrix template either.

In practical data-mining applications, n, the number of records, ma
be as large as 10^{12} and m, the number of attributes, as large as 10^4. Thes
values are growing all the time as datasets themselves get larger, and as be
ter algorithms and hardware make it cost-effective to attack large dataset
directly.

To illustrate the techniques we are discussing we will use the following $11 \times$
matrix:

$$A = \begin{bmatrix} 1 & 2 & 3 & 4 & 5 & 6 & 7 & 8 \\ 3 & 4 & 4 & 5 & 5 & 6 & 7 & 9 \\ 1 & 8 & 2 & 7 & 3 & 6 & 4 & 5 \\ 9 & 8 & 7 & 6 & 5 & 4 & 3 & 2 \\ 9 & 4 & 8 & 3 & 7 & 2 & 6 & 1 \\ 2 & 3 & 2 & 4 & 2 & 5 & 2 & 6 \\ 3 & 4 & 3 & 4 & 4 & 3 & 4 & 3 \\ 3 & 2 & 4 & 3 & 2 & 4 & 3 & 2 \\ 5 & 5 & 4 & 4 & 6 & 6 & 2 & 2 \\ 2 & 3 & 6 & 5 & 4 & 6 & 7 & 2 \\ 1 & 6 & 5 & 3 & 8 & 2 & 3 & 9 \end{bmatrix}$$

If we think of each row as describing the properties of an object, and th.
columns as describing a set of attributes, then we can see that objects 1 an.
2 (and to a lesser extent 3) have small values for the first few attribute.
increasing for the later attributes; objects 4 and 5 have the opposite patter
– mostly large values for the first few attributes and smaller ones for th.
later attributes; while objects 6 to 11 have moderate values for most of th.
attributes. Of course, we can only pick out such properties by inspection whe
the matrix is relatively small, and when the rows have been arranged to mak.
it easy.

We will use this matrix as an example throughout the book. A Matla.
script used to generate all of the data and figures based on this matrix can t.
found in Appendix A.

1.2 Data-mining techniques

Many kinds of analysis of data are possible, but there are four main kinds:

1. *Prediction*, producing an appropriate label or categorization for new
   objects, given their attributes, using information gleaned from the rela.
   tionship between attribute values *and* labels of a set of example object.

2. *Clustering*, gathering objects into groups so that the objects within
   group are somehow similar, but the groups are somehow dissimilar.

3. *Finding outliers*, deciding which objects in a given dataset are the m
   unusual.

4. *Finding local patterns*, finding small subsets of the objects that h
   strong relationships among themselves.

1.2.1   Prediction

In *prediction*, the goal is to predict, for a new record or object, the va
of one of the attributes (the 'target attribute') based on the values of
other attributes. The relationship between the target attribute and the ot
attributes is learned from a set of data in which the target attribute is
ready known (the 'training data'). The training data captures an empiri
dependency between the ordinary attributes and the target attribute;
data-mining technique builds an explicit model of the observed dependen
This explicit model can then be used to generate a prediction of the tar,
attribute from the values of the other attributes for new, never before se
records. When the target values are categorical, that is chosen from so
fixed set of possibilities such as predicting whether or not a prospective b
rower should be given a mortgage, prediction is called *classification*. Wl
the target values are numerical, for example predicting the size of mortg
a prospective borrower should be allowed, prediction is called *regression*.

Each data-mining technique assumes a different form for the expl
prediction model, that is a different structure and complexity of the dep
dencies among the attributes. The quality of a model can be assessed usin
test set, a subset of the data for which the correct target attribute values
known, but which was not used as part of the training data. The accuracy
predictions on the test set is an indication of how the model will perform
new data records, and so how well it has captured the dependencies amo
the attributes.

The simplest prediction model is the *decision tree*, a technique rela
to the well-known game of Twenty Questions. A decision tree is (usually
binary tree, with an inequality test on one of the attributes at each inter
node, and a target attribute value associated with each leaf. The tar
attribute must be *categorical*, that is with values from a fixed set. Wl
a new object is to be classified, it begins at the root node. If its attrib
values satisfy the inequality there, then it passes down (say) the left brar
otherwise it passes down the right branch. The same process is repeated
each internal node, so the object eventually ends up at one of the leaves.
predicted target attribute value for the new object is the one associated w
that leaf.

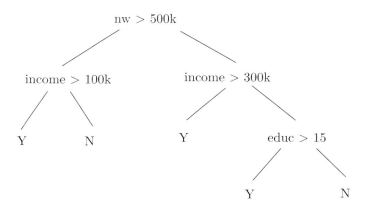

Figure 1.1. *Decision tree to decide individuals who are goo prospects for luxury goods.*

Suppose a company wants to decide who might be interested in buyin a luxury product such as an expensive watch. It has access to the net worth income, and years of education of customers who have previously bought th product and wants to decide which new individuals should be approached t buy the product. Figure 1.1 shows a decision tree that might be constructe based on existing customers. The internal nodes represent decisions base on the available attributes of new customers, with the convention that th branch to the left describes what to do if the inequality is satisfied. Th leaves are labelled with the class labels, in this case 'yes' if the customer good prospect and 'no' if the customer is not. So, for example, a potentia customer whose net worth is below \$500,000 but whose income is more tha \$300,000 is considered a good prospect.

The process of *constructing* a decision tree from training data is mor complicated. Consider the process of deciding which inequality to choose fo the root node. This requires first selecting the attribute that will be used and second selecting the boundary value for the inequality that will define th separation between the two descendant nodes. Given the training data, eac attribute is examined in turn and the one that provides the most 'discrimina tion' is selected. There are a number of ways of instantiating 'discrimination for example *information gain*, or *gini index*, details of which can be found i standard data-mining texts. The value of that attribute that is most 'discrim inating' is selected. Again, the details can be found in standard texts. Th process of growing the tree stops when the training data objects associate with each leaf are sufficiently 'pure', that is they mostly have the same valu for their target attribute.

The tree structure and the construction process are slightly different attributes can be categorical, that is have values chosen from a fixed set o

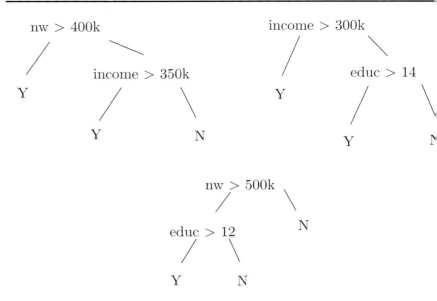

Figure 1.2. *Random forest of three decision trees, each trained*
two attributes.

possibilities. If a categorical attribute is chosen as most discriminating, t
the tree is no longer binary – it has one descendant for each possible va
that the attribute can take, and the test is a match against these values.
decision about which attribute is most discriminatory is done in essenti
the same way, but there are one or two details to consider. For exam
a categorical attribute with many possible values looks more discriminat
than one with few possible values, but this is not necessarily a reason to pr
it.

Another prediction technique based on decision trees is *random fore*
Instead of growing a single decision tree from the training data, multi
decision trees are grown. As each tree is being grown, the choice of
best attribute on which to split at each internal node is made from amon
randomly-chosen, fixed size subset of the attributes. The global prediction
derived from the predictions of each tree by *voting* – the target attribute va
with the largest number of votes wins. Random forests are effective predict
because both the construction mechanism and the use of voting cancels
variance among the individual trees – producing a better global predictio

A set of possible decision trees for predicting prospects for luxury pr
ucts is shown in Figure 1.2. Each of the decision trees is built from a subse
the available attributes, in this case two of the three. Because only a sub
of the data is being considered as each tree is built, attributes can be cho
in different orders, and the inequalities can be different. In this case, an i

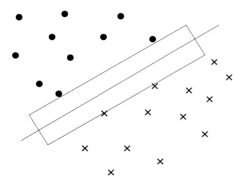

Figure 1.3. *Thickest block separating objects of two classes (circle and crosses), with midline defining the boundary between the classes.*

vidual whose net worth is $450,000, whose income is $250,000, and who ha 15 years of education will be regarded as a good prospect. The first two tree classify the individual as a good prospect, while the third does not. Howeve the overall vote is two to one, so the global classification is 'good prospect Notice that the amount of agreement among the trees also provides an est mate of overall confidence in the prediction. An individual with net wort $450,000, income of $350,000 and 15 years of education would be considered good prospect with greater confidence because the vote for this classificatio is three to zero.

A third prediction technique is *support vector machines* (SVMs). Th technique is based on a geometric view of the data and, in its simplest form predicts only two different target attribute values. A data record with m at tribute values can be thought of as a point in m-dimensional space, by treatin each of the attribute values as a coordinate in one dimension. Support vecto machines classify objects by finding the best hyperplane that separates th points corresponding to objects of the two classes. It uses three importan ideas. First, the best separator of two sets of points is the midline of the thick est plank or block that can be inserted between them; this allows the problen of finding the best separator to be expressed as a quadratic minimizatio problem.

Figure 1.3 shows an example of objects from a dataset with two at tributes, plotted in two-dimensional space. The thickest block that can fi between the objects of one class (circles) and the objects of the other clas (crosses) is shown; its midline, also shown, is the best boundary betwee the classes. Notice that two circles and two crosses touch the separatin block. These are the support vectors, and the orientation and placemen of the boundary depends only on them – the other objects are irrelevant i determining the best way to separate the two classes.

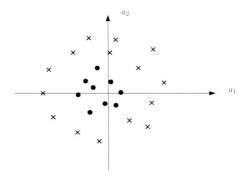

Figure 1.4. *Two classes that cannot be linearly separated.*

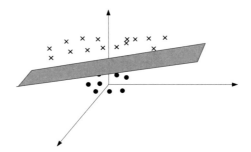

Figure 1.5. *The two classes can now be linearly separated in third dimension, created by adding a new attribute $abs(a_1) + abs(a_2)$.*

Second, if the two classes are not well separated in the space span by the attributes, they may be better separated in a higher-dimensional sp spanned both by the original attributes and new attributes that are combi tions of the original attributes.

Figure 1.4 shows a situation where the objects in the two classes can be linearly separated. However, if we add a new attribute, $abs(a_1) + abs($ to the dataset, then those objects that are far from the origin in the t dimensional plot (crosses) will now all be far from the origin in the th dimension too; while those objects close to the origin (circles) will rem close to the origin in the third dimension. A plane inserted roughly para to dimensions 1 and 2 will now separate the two classes linearly, as showr Figure 1.5. A new object with values for attributes a_1 and a_2 can be mapp into the three dimensional space by computing a value for its third attribu and seeing which side of the plane the resulting point lies on.

Third, the form of the minimization requires only inner products of objects and their attributes; with some care, the combinations of attribu required for a higher-dimensional space need not ever be actually compu because their inner products can be computed directly from the original attributes.

The SVM technique can be extended to allow some objects to be on the 'wrong' side of the separator, with a penalty; and to allow different forms of combinations of the original attributes. Although SVMs compute only two-class separators, they can be extended to multiclass problems by building separators pairwise for each pair of classes, and then combining the resulting classifications.

Many other prediction techniques are known, but random forests and support vector machines are two of the most effective.

1.2.2   Clustering

In *clustering*, the goal is to understand the macroscopic structure and relationships among the objects by considering the ways in which they are similar and dissimilar. In many datasets, the distribution of objects with respect to some similarity relationship is not uniform, so that some of the objects resemble each other more closely than average. Such a subset is called a cluster. In a good clustering, objects from different clusters should resemble each other less than average. For any particular dataset, there are many ways to compare objects, so a clustering always implicitly contains some assumption about the meaning of similarity.

Clustering techniques can be divided into three kinds: those based on distances among objects in the geometrical sense described above (clusters are objects that are unusually close to each other); those based on density of objects (clusters are regions where objects are unusually common); or those based on probability distributions (clusters are sets of objects that fit an expected distribution well). These are called *distance-based*, *density-based*, and *distribution-based* clusterings, respectively.

Clustering techniques can also be distinguished by whether they carve up the objects into disjoint clusters at a single level (*partitional clustering*) or give a complete hierarchical description of how objects are similar to each other (*hierarchical clustering*), using a dendrogram. As well, some clustering techniques need to be told how many clusters to look for, while others will try to infer how many are present.

The simplest geometrical clustering technique is *k-means*. Given a data set considered as a set of points in m-dimensional space, a set of k cluster centers are chosen at random. Each point in the dataset is allocated to the nearest cluster center. The centroid of each of these allocated sets of points is computed, and these centroids become the new cluster centers. The process is repeated until the cluster centers do not change. Each set of points

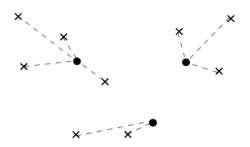

Figure 1.6. *Initialization of the k-means algorithm, with obje denoted by crosses, and k initial cluster centers denoted by circles. The das lines indicate which cluster center is closest to each object.*

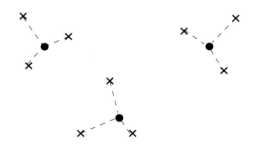

Figure 1.7. *Second round of the k-means algorithm. One object moved from one cluster to another, and all objects are closer to their cer than in the previous round.*

allocated to (closest to) a cluster center is one cluster in the data. Beca k is a parameter to the algorithm, the number of clusters must be known guessed beforehand.

Figures 1.6 and 1.7 show a small example in two dimensions. The cros represent data points. If the cluster centers (circles) are placed as shown Figure 1.6, then each object is allocated to its nearest cluster center. T relationship is shown by dashed lines. After this initial, random, allocati each cluster center is moved to the centroid of the objects that belong to as shown in Figure 1.7. Since the centers have moved, some objects will closer to a different center – one point has been reallocated in Figure The allocations of objects to new cluster centers is again shown by the das lines. It is clear that the allocation of objects to clusters will not cha further, although the cluster centers will move slightly in subsequent rou of the algorithm.

The k-means algorithm is simple and fast to compute. A poor cho of the initial cluster centers can lead to a poor clustering, so it is comm to repeat the algorithm several times with different centers and choose clustering that is the best. It is also possible to make cleverer choices of the initial cluster centers, for example by choosing them from among the objects or by calculating some simple distributional information from the data and using that to make better initial choices.

Typical density-based partitional clustering algorithms choose an object at random to be a potential cluster 'center' and then examine its neighborhood. Objects that are sufficiently close are added to the cluster, and then their neighbors are considered, in turn. This process continues until no further points are close enough to be added. If enough points have been found, that the potential cluster is large enough, then it becomes one of the clusters and its members are removed from further consideration. The process is repeated until no new clusters can be found. Some objects may not be allocated to any cluster because there are not enough other objects near them – this can be either a disadvantage or advantage, depending on the problem domain.

The best known distribution-based clustering technique is *Expectation Maximization* (EM). Instead of assuming that each object is a member of exactly one cluster, the EM approach assumes that clusters are well-represented by probability density functions, that is regions with a center and some variability around that center, and objects belong to each cluster with some probabilities. Suppose that the dataset contains two clusters, and we have some understanding of the shape of the clusters. For example, they may be multidimensional Gaussians, so we are hypothesizing that the data is well described as a mixture of Gaussians. There are several missing values in this scenario: we do not know the parameters of the distributions, and we do not know the probability that each object is in cluster 1. The EM algorithm computes these missing values in a locally optimal way.

Initially, all of the missing values are set randomly. In the Expectation (E) step, the expected likelihood of the entire dataset with these missing values filled in is determined. In the Maximization (M) step, the missing values are recomputed by maximizing the function from the previous step. These new values are used for a new E step, and then M step, the process continuing until it converges. The EM algorithm essentially guesses values for those that are missing, uses the dataset to measure how well these values 'fit', and then re-estimates new values that will be better. Like k-means, EM can converge to a local maximum, so it may need to be run several times with different initial settings for the missing values.

Figure 1.8 shows an initial configuration for the EM algorithm, using the same data points as in the k-means example. The ellipses are equi-probable contours of 2-dimensional Gaussian distributions. The point labelled A has some probability of belonging to the bottom distribution, a lower probability of belonging to the top, left distribution, and a much smaller probability of belonging to the top, right distribution. In the subsequent round, shown in Figure 1.9, the parameters of the bottom distribution have changed to make

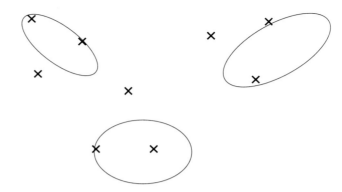

Figure 1.8. *Initial random 2-dimensional Gaussian distributio each shown by a probability contour. The data points are shown as crosse.*

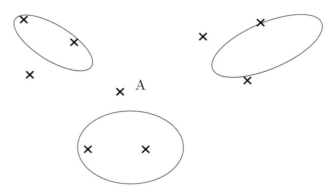

Figure 1.9. *Second round of the EM algorithm. All three distra tions have changed their parameters, and so their contours, to better expl the objects, for example, object A.*

it slightly wider, and hence increasing the probability that A belongs to while the other two distributions have changed slightly to make it less lik that A belongs to them. Of course, this is a gross simplification, since al the objects affect the parameters of all of the distributions, but it gives flavor of the algorithm.

Hierarchical clustering algorithms are usually bottom-up, and begin treating each object as a cluster of size 1. The two nearest clusters are join to form a cluster of size 2. The two nearest remaining clusters are join and so on, until there is only a single cluster containing all of the obje There are several plausible ways to measure the distance between two cl ters that contain more than one object: the distance between their near members, the distance between their centroids, the distance between th

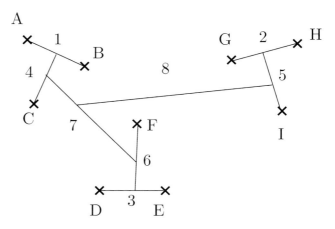

Figure 1.10. *Hierarchical clustering of objects based on proximity i two dimensions. The edges are numbered by the sequence in which they we created to join the clusters at their two ends.*

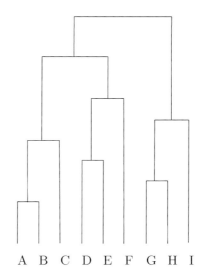

Figure 1.11. *Dendrogram resulting from the hierarchical clusterin Any horizontal cut produces a clustering; the lower the cut, the more cluste there are.*

furthest members, and several even more complex measures. Hierarchic clustering can also be done top-down, beginning with a partitioning of th data into two clusters, then continuing to find the next best partition an so on. However, there are many possible partitions to consider, so top-dow partitioning tends to be expensive.

Figure 1.10 shows a hierarchical clustering of our example set of obje in two dimensions. The edges are numbered in the order in which they mi, be created. Objects A and B are closest, so they are joined first, becomin cluster of size 2 whose position is regarded as the centroid of the two obje All of the objects are examined again, and the two closest, G and H, joined to become a cluster. On the third round, objects D and E are join On the fourth round, the two nearest clusters are the one containing A a B, and the one containing only C, so these clusters are joined to produc cluster containing A, B, and C, and represented by their centroid. The proc continues until there is only a single cluster. Figure 1.11 shows a dendrogr that records this clustering structure. The lower each horizontal line, earlier the two subclusters were joined. A cut across the dendrogram at a level produces a clustering of the data; the lower the cut, the more clust there will be.

1.2.3   Finding outliers

In *finding outliers*, the goal is to find those objects that are most unusu rather than to understand the primary structure and relationships among objects. For example, detecting credit card fraud requires finding transacti that are sufficiently unusual, since these are likely to be misuse of a card. in clustering, there must be some implicit assumption about the meaning similarity (or dissimilarity).

Not many techniques for finding outliers directly are known. One-cl support vector machines try to capture the main structure of the data fitting a distribution such as a multidimensional Gaussian to it. Those jects on or just outside the boundary are treated as outliers. Although sc successes with this technique have been reported in the literature, it seems be extremely sensitive to the parameter that describes how tightly the m data is to be wrapped.

Density-based clustering techniques can be used to detect outliers, si these are likely to be those objects that are not allocated to any clus Hierarchical algorithms can also detect outliers as they are likely to be sir objects or very small clusters that are joined to the dendrogram only at lev close to the root.

1.2.4   Finding local patterns

In *finding local patterns*, the goal is to understand the structure and relati ships among some small subset(s) of the objects, rather than understand the global structure. For example, in investigations of money laundering, primary goal may be to find instances of a cash business connected to ba

accounts with many transactions just under $10,000. The many other possib[
relationships among objects are of less interest.

The most common technique for finding local patterns is *associatio*
rules, which have been successful at understanding so-called market-bask(
data, groupings of objects bought at the same time, either in bricks-and
mortar stores, or online. Suppose we have a dataset in which each row rep
resents a set of objects purchased at one time. We would like to learn, fo
example, which objects are often purchased together.

Objects that are purchased together only 1 in 10,000 times probab[
do not have much to tell us. So it is usual to consider only sets of object
that occur together in a row more than some fraction of the time, called th
support. Finding such *frequent sets* of objects that occur together depenc
on the observation that a set of k objects can be frequent only if all of i
subsets are also frequent. This leads to the *levelwise* or *a priori* algorithm
compute all pairs of objects that are frequent; from these pairs compute on[
those triples that could be frequent (for example, if AB, AC, and BC ar
all frequent then ABC *might* be frequent), and check which of these triple
actually are frequent, discarding the rest. Repeat by combining frequer
triples into potentially frequent quadruples of objects; and so on. It become
harder and harder to find sets of objects that might be frequent as the set
get larger, so the algorithm runs quickly after it passes the first step – ther
are typically many potentially frequent pairs.

Each set of frequent objects can be converted into a series of rules b
taking one object at a time, and making it the left-hand side of a rule whos
right-hand side is the remaining objects. For example, if ABC is a frequer
set, then three rules: A → BC, B → AC, and C → AB can be derived from i
The predictive power of these rules depends on how often the left-hand sid
predicts the presence of the right-hand side objects in the same purchase,
quantity called the *confidence* of the rule. Confidences are easily compute
from frequencies. If the frequency of the set ABC is 1000 and the frequenc
of the set BC is 500, then the confidence of the rule A → BC is 0.5.

The problem with local rules is that it is often difficult to decide how t
act on the information they reveal. For example, if customers who purchas
item A also purchase item B, should As and Bs be placed together on a she
to remind customers to buy both? Or should they be placed at opposite end
of the store to force customers to walk past many other items that might b
bought on impulse? Or something else?

1.3   Why use matrix decompositions?

The standard data-mining techniques described above work well with man
common datasets. However, the datasets that arise in settings such as sci

ence, engineering, medicine, economics, and politics are often complex in wa
that make mainstream data-mining techniques ineffective. Two properties
complex datasets that make straightforward data mining problematic are

1. Data comes from multiple processes. Each entry in the dataset is usua
   not the result of a single, discrete property or action of the object w
   which it is associated; rather it is the combination of values that ha
   arisen from different processes, and have been combined to produce
   single value captured in the dataset.

2. Data has multiple causes. The relationships among the attributes, a
   between each attribute and the target attribute, are subtle, and so
   attributes are predictive only for some records.

For these complex datasets, more powerful analysis techniques are
quired. Matrix decompositions can sometimes provide this more power
analysis; at other times, they can provide ways to produce cleaner data wh
mainstream techniques may then be able to use.

1.3.1   Data that comes from multiple processes

One way in which the values recorded in a dataset can vary from their 'tr
values is when the collection process introduces noise. Most data collec
in the real world comes from measuring devices, and these almost alw
introduce noise into the data. Noise can also be introduced into data in ot
ways; for example, if people are asked their opinions, they often respond
inaccurate ways, especially if they are asked about sensitive topics.

Individuals are also quite variable over short time scales. Even if
topic is not sensitive, a response today may be substantially different fr
what it was yesterday, or will be tomorrow, so the data collected is reall
sample of a much more complex set of possible responses.

Unsurprisingly, the presence of noise can distort predictive models
clusterings built from the data. Not so obviously, the distortions introduced
noise can be much larger than expected from the magnitude of the noise its
The support vector technique tells us that the best separator is the midlin
the thickest rectangle that can be inserted between points representing obje
from the two classes. Only some of the objects, those that make contact w
this rectangle, actually alter its placement – such points are called supp
vectors, and they are typically a small fraction of the objects. If only
or two of these objects are moved by a small amount because of noise
can make a large difference to the angle (and thickness) of the separat
rectangle, and so to the boundary between the two classes.

Analyzing a dataset without considering how much its values may have been distorted by noise can lead to poor results. Matrix decompositions provide several ways to investigate datasets for the presence of noise, and also allow it to be removed.

Even when noise is not present in a dataset, for example because it has been collected by some automatic and digital process, it is still possible for the values in the dataset to represent a merging of data from different underlying processes. Consider the collection of data about stars using a telescope. The measured intensity of a particular star at a particular wavelength represents the sum of terms that represent: the actual intensity of the star at that wavelength, the gravitational force on the light due to other stars and galaxies, the effects of the atmosphere on the light as it passed from space to the telescope, and properties of the telescope itself. Treating such a measured value as if it represented only the actual intensity of the star is bound to create inaccurate models.

This situation could be thought of as noise corrupting an underlying signal, but it is better to regard this as four different processes whose effects are superimposed in the observed data. After all, which process represents noise depends on the goals of the observation; it might be that the goal is to develop a new detection system to observe stealth aircraft which give themselves away by the way they distort the light from stars. With this goal, the process that might previously have been considered noise is now signal.

In general, many datasets cannot capture significant data without also capturing other kinds of data that blur the structures of interest. Trying to model the data without awareness of these other processes' contributions leads to weak models. In the astronomical example above, the four processes could plausibly be assumed to be statistically independent, which makes it easier to separate them, but even this is not usually the case. The contributions of each of the processes are intertwined, and separating them is not easy.

1.3.2 Data that has multiple causes

It is attractive to think that, in any given dataset, there is a fixed relationship between a particular attribute and the target attribute; attribute a_1 is strongly predictive, attribute a_2 is not very predictive, and so on. Indeed, there are algorithms for attribute selection that assume this kind of relationship and aim to select those attributes that are most useful, so that the others can be discarded, and the analysis simplified.

It is sometimes the case that some attributes in a dataset are almost useless in determining its underlying structure. This happens partly because datasets are collected for many reasons, and subsequent data mining is an extra win. It also happens because it is usually hard to tell, in advance

which attributes will be the most revealing, and so it makes sense to coll
as many as possible.

However, in many real-world datasets, the dependence of the over
structure on individual attributes is more complex. It is often the case tl
attribute a_1 is predictive for objects from class 1, whereas attribute a_3
predictive for objects from class 2, and attribute a_5 is predictive only
some other objects from class 2. Attribute selection, in the usual sense, d
not help for such datasets. Discarding an attribute means discarding use
information; on the other hand, any particular attribute may be useful
only some of the objects and useless, or even misleading, for other objects

A simple example may help to clarify this issue. Suppose you are as
what personality traits make you like someone, and what personality tra
make you dislike someone. There will probably be some traits that are on o
of your lists, while their opposites are on the other list. These correspond
attributes that are uniformly predictive of whether you will or will not l
someone. However, there are likely to be some traits that appear on one
your lists, but whose opposites do not appear on the other. For example, y
might say that you dislike people who talk a lot. It does not necessarily foll
that you will like quiet people. Traits that appear on only one list correspo
to attributes that are predictive for one class, but have no predictive pov
for the other class.

Consider the simple scenario shown in Figure 1.12. Here objects fr
Class 1 (circles) are easily determined because they contain only a limi
range of values for attribute a_1; their values for attribute a_2 are widely o
tributed. In the same way, objects from Class 2 (crosses) are easily determi
by a limited range of values for attribute a_2; but their values for attribute
are widely distributed. Both predictors and clusterings of this data will ins
a boundary between the classes roughly as shown by the line.

However, it is easy to see from the Figure that the precise placem
of the boundary depends on the exact positions of the objects in the
right hand corner! These objects are the least typical objects of either cla
and so the least reliable on which to base decisions. Yet these objects, t
are least characteristic, and perhaps have attribute values that are least to
trusted, are those that contribute most strongly to the model being built. T
situation is typical of many datasets, and illustrates the pitfalls of assum
that attributes, rather than attribute values, correlate with classes.

1.3.3   What are matrix decompositions used for?

Matrix decompositions have two main roles in data analysis. The firs
that they are able to tease apart the different processes that have usua
been captured by the dataset. The effects of processes that are irrelev

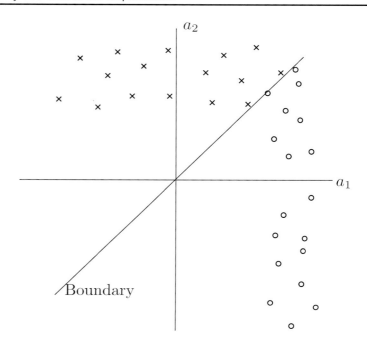

Figure 1.12. *Typical data distribution for a simple two-attribute two-class problem.*

to the task at hand, perhaps noise or some other unavoidable processes that are intertwined with the process of interest, can be removed from the data. This enables subsequent modelling using mainstream data mining to produce better results. This role might be called *data cleaning*.

The second role for matrix decompositions is to cluster the objects (or attributes) of a dataset, either directly in ways that derive from the matrix decomposition or in some standard way.

Matrix decompositions also allow other forms of analysis, for example experimenting with the importance of a critical object or attribute, and forcing representations of the data in terms of a small number of substructures. We will see many of these in action in subsequent chapters.

Notes

The scientific method is discussed in Popper's book [91].

Some standard data-mining texts are those by: Tan, Steinbach and Kumar [110], Kantardzic [64], and Dunham [39]. A book written from a more statistical perspective is Hand, Mannila, and Smyth [50]. Although there w a number of good survey papers about data mining, most of these are n quite old.

Decision trees were developed by Quinlan [92–94] and independently Breiman [18]. Random Forests were also developed by Breiman [17]. Supp Vector Machines, as they are used today, were developed by Cortes and Vap [30], and extended by a large number of others [15, 22, 31, 37].

The k-means algorithm was developed by MacQueen [83], and has b used in hundreds of different contexts. It is often used as a first algorit to cluster a dataset. However, its use of Euclidean distance assumes t clusters are naturally spherical, and this assumption almost never holds practice. An example of a density-based clustering algorithm is DBSCAN [The Expectation-Maximization algorithm was developed by Dempster [Hierarchical clustering is a simple idea that should probably be credited Linnaeus's classification of species. However, the approach is usually credi to Johnson [61]. Association rules and algorithms to compute them w developed in a series of papers by Agrawal and others [5, 6].

Chapter 2

Matrix decompositions

Matrix decompositions have been used for almost a century for data analysis and a large set of different decompositions are known. The most important ones are:

- Singular Value Decomposition (SVD), and its close relation, Principal Component Analysis (PCA);

- SemiDiscrete Decomposition (SDD);

- Independent Component Analysis (ICA);

- Non-Negative Matrix Factorization (NNMF);

Some of these are really families of related decompositions; there are also number of variants and extensions, and we will briefly discuss some of them as well.

2.1   Definition

Recall that we consider a dataset as a matrix, with n rows, each of which represents an object, and m columns, each of which represents an attribute. The ijth entry of a dataset matrix is the value of attribute j for object i. Each family of matrix decompositions is a way of expressing a dataset matrix

---

[1]In some applications it is more natural to use the rows of A for the attributes and the columns for objects. This doesn't change anything since the transpose of a matrix product is the product of the transposes in the reverse order. However, it does make reading the literature confusing.

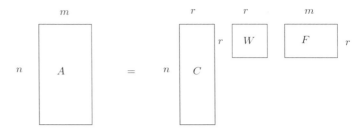

Figure 2.1. *A basic matrix decomposition.*

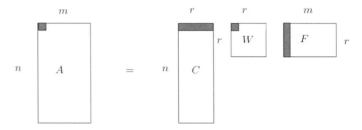

Figure 2.2. *Each element of A is expressed as a product of a row*
C, an element of W, and a column of F.

A, as the product of a set of new matrices, usually simpler in some way, t.
shed light on the structures or relationships implicit in A. Different mat
decompositions reveal different kinds of underlying structure.

More formally, a matrix decomposition can be described by an equat
of this form
$$A = C \, W \, F \tag{2}$$
where the sizes of the matrices are as follows: A is $n \times m$ (and we assu
for simplicity that $n > m$; in practice $n \gg m$); C is $n \times r$ for some r tha
usually smaller than m; W is $r \times r$, and F is $r \times m$; Figure 2.1 illustrate
matrix decomposition.

From this equation, an element of A, say a_{11}, arises from the multi
cation of the first row of C, the top left element of W, and the first colu
of F, as shown in Figure 2.2. If we think of the rows of F as parts or piec
then the product $W \, F$ weights each of the rows by the corresponding diago
element of W. The matrix C then takes something from each part and co
bines them in a weighted way. Hence each entry of A is a kind of combinat
of parts from F, combined in ways described by C and W.

The matrix C has the same number of rows as A. Each row of C give
different view of the object described by the corresponding row of A. In ot
words, the ith row of C provides r pieces of information that together ar

new view of the ith object; while A provides m pieces of information abou
the ith object.

The matrix F has the same number of columns as A. Each colum
of F gives a different view of the attribute described by the correspondin
column of A, in terms of r pieces of information, rather than the n pieces ϵ
information in A.

The role of r is to force a representation for the data that is more compac
than its original form. Choosing $r = m$ still gives a sensible decompositio
but it is usually the case that r is chosen to be smaller than m. We are in
plicitly assuming that a more compact representation will capture underlyin
or latent regularities in the data that might be obscured by the form in whic
the data is found in A, usually because A expresses the data in a way th
contains redundancies.

The particular dataset, A, being studied is always considered to be
sample from a larger set of data that *could* have been collected. The us
of a limited representational form prevents the matrix decomposition fro
overfitting the data, that is learning the precise properties of *this* particul
dataset, rather than the properties of the larger system from which it came

The matrix W has entries that reflect connections among the differer
latent or implicit regularities or latent factors – the ijth entry provides th
relationship between the latent factor captured by the ith column of C (
kind of latent attribute) and latent factor captured by the jth row of F (
kind of latent object). For us, W will always be a diagonal matrix (that is, i
off-diagonal elements are zero), in which case the latent factors for the object
and attributes are the same, and each entry can be interpreted as providin
information about the relative importance of each underlying factor. Som
decompositions do not create this middle matrix, but we can always imagin
that it is there as the $r \times r$ identity matrix.

Usually r is smaller, often much smaller, than m, but a few matri
decompositions allow $r > m$. In this case, the underlying factors must som
how be of a particular simple kind, so that the matrix decomposition is sti
forced to discover a compact representation. For example, as we shall see, th
SemiDiscrete Decomposition allows $r > m$, but the entries of the C and A
matrices are restricted to be only $-1, 0$, or $+1$.

We will consider many different kinds of matrix decompositions. Thes
differ from each other in the assumptions they make about the kind of ur
derlying structure that can be present in the data. In practice, this mean
that different matrix decompositions have different requirements for the er
tries of the matrices into which the dataset is decomposed, different relatior
ships among the rows and columns, and different algorithms to compute eac
decomposition. Nevertheless, there are deep connections among the matri

decompositions we will consider. Most can be expressed as constrained timization problems; and all are a form of Expectation-Maximization w stringent requirements on the distributions assumed.

Symmetry between objects and attributes

There is always a kind of symmetry between the objects and the attributes i dataset because the matrix decomposition on the objects, as we have describ it, can also be turned into a matrix decomposition on the attributes. For

$$A = C \, W \, F$$

then

$$A' = F' \, W' \, C'$$

The dash indicates the transpose of the matrix, that is the matrix obtain by flipping the original matrix across its main diagonal, making rows i columns, and columns into rows. A' reverses the roles of objects and tributes, so the attributes are now the rows. On the right-hand side, F' pl the role originally played by C, and C' plays the role originally played by For any matrix decomposition, whatever can be done with the objects in dataset can also be done with the attributes, and *vice versa*.

Normalization

Because matrix decompositions are numerical computations, the magnitu of the values in different columns (different attributes) must be comparal or else the large magnitudes will have a greater influence on the result tl the smaller ones. How this is done, however, requires some care because amounts to making assumptions, perhaps quite strong ones, about the da Although matrix decompositions are usually characterized as non-paramet methods, the choice of normalization is really a parameter.

One standard way to adjust attribute values is to subtract the mean fr the entries in each column, which centers the values around zero; and then vide each entry in each column by the standard deviation of the column me This makes the values in different columns roughly similar in magnitude, implicitly assumes that the values of each attribute are normally distribut We will discuss normalization in detail for each matrix decomposition.

When it is not clear how to normalize values in the dataset, as for ample when the distribution of values is very different for different attribut it can often be useful to replace the values in each column by their ranks. common way to do this is to use the *Spearman rank*. The values are nu bered in increasing order, except that when there are ties, the rank associa with the tied elements is the average of the ranks that those elements would have had if they had been different. Suppose that the original values are, say 1,4,2,3,2,4,2. Sorting these into order we get 1,2,2,2,3,4,4 and the corresponding ranks are 1, 3 (= (2+3+4)/3), 3, 3, 5, 6.5 (= (6+7)/2), and 6.5. Each column of the dataset contains the same number of values, so the magnitudes in the different columns are roughly the same.

Degenerate decompositions

Many decompositions, in their simple forms, can be degenerate. Given an invertible $m \times m$ matrix X, it is often possible to insert $X\,X^{-1}$ in the right-hand side of a decomposition, rearrange, and get a new right-hand side that is another example of the same decomposition. If

$$A \;=\; C\,F$$

then

$$A \;=\; C\,(X\,X^{-1})\,F \;=\; (C\,X)\,(X^{-1}\,F)$$

and the parenthesized terms on the right-hand side are a new C and a new F and so a different decomposition of A. Most matrix decompositions impose some further condition to specify which, of all these related decompositions, is 'the' decomposition.

Correlation matrices

Given a matrix A, we can form the matrices AA' and $A'A$, where the dash indicates the transpose of A. The matrix AA' is the *correlation matrix* of the objects. The magnitude of the ijth entry indicates the amount of correlation between the ith and the jth object. Similarly, the matrix $A'A$ is the *correlation matrix* of the attributes, and its entries indicate the amount of correlation between pairs of attributes. Both of these matrices are symmetric.

The correlation matrices can also be decomposed, and the resulting matrices analyzed to give new insights into the structures present in the data. However, this is often not as helpful as it seems, for three reasons. First, the correlation matrices are $n \times n$ and $m \times m$ respectively, so that at least the first can be very large. Second, calculating a decomposition for such a matrix can often be difficult because of numerical instability. Third, each decomposition of a correlation matrix provides information about the structure of the objects or about the structure of the attributes, but not both at once. This can lose information implicit in their interactions.

2.2   Interpreting decompositions

Equation 2.1 explains what a matrix decomposition is, but it does not expl
how to compute one, or how such a decomposition can reveal the structu
implicit in a dataset. The computation of a matrix decomposition is straig
forward; software to compute each one is readily available, and understandi
in a deep way, how the algorithms work is not necessary to be able to interp
the results.

Seeing how a matrix decomposition reveals structure in a dataset is m
complicated. Each decomposition reveals a different kind of implicit struct
and, for each decomposition, there are four different, although related, ways
interpret the result. Hence, for each dataset, there are many possible aven
for exploration.

Each decomposition allows the following four interpretations:

- A factor interpretation. Here the underlying assumption is that the r
  of F represent r underlying or hidden factors with inherent significan
  The objects in the observed data, A, are the result of *mixing* th
  underlying factors in different proportions given by the entries of e;
  row of C.

- A geometric interpretation. Here the underlying assumption is t
  the rows of A can be interpreted as coordinates in an m-dimensio
  space. After decomposition, each object is described by a new set
  coordinates, the entries of the corresponding rows of C with respect
  a set of axes given by the r rows of F.

- A component interpretation. Here the underlying assumption is t
  each entry in the dataset is a blend of values from different proces
  that contributed to the dataset. The component interpretation natura
  allows such contributions to be separated.

- A graph interpretation. Here nodes are associated with each of
  objects and the attributes, with edges joining nodes of one kind to no
  of the other kind, weighted by the matrix entries. If these weights
  regarded as permeability of edges, then objects that are similar to e
  other are nodes that have many 'easy' paths between them.

These four interpretations might be called the *hidden factors* model, the *h
den clusters* model, the *hidden processes* model, and the *hidden connecti*
model, respectively. The four interpretations are mathematically equivale
but provide different views of the structure hidden within the matrix A.
each particular dataset and application domain, one or other of these interp
tations will often seem more natural, but it is usually instructive to consi
all four.

2.2.1 Factor interpretation – hidden sources

In this interpretation, the attributes that are captured in the dataset A a
regarded as mixtures (somehow) of some attributes that have more dire
significance. For example, if sound signals are captured by microphones i
a number of places in a noisy room, then the amplitudes and frequencies a
these locations are the measured attributes. The real attributes of interes
might be what the speakers in the room are saying, that is the amplitudes an
frequencies of the sounds each speaker produces. The measured attribute
of course, depend on the real attributes, but in ways that also depend o
where the microphones are placed and the shapes and textures of objects i
the room. This is often called the *Blind Source Separation* problem, and it
ubiquitous in signal processing.

Data analysis in the humanities and social sciences often uses this in
terpretation as well. Data collected by surveys, for example, often describe
superficial properties or actions, while the goal of the survey is to discover th
deeper drivers or motivators that explain or cause the superficial propertie
or actions.

For example, data collected about athletes might contain informatio
about their heights, weights, femur length, and shoe size. These properties ar
probably highly correlated and reflect a latent property, 'body size'. Reducin
these overt factors to a single, latent factor might make it easier to see wh
some athletes are successful and others are not. A decomposition shows ho
this latent factor contributes to the observed factors via the entries in C; tha
is, the value of height for a given athlete is a scalar multiple of the 'body siz
factor, the value for weight is a different scalar multiple, and so on. Thes
scalars appear along the row of C corresponding to the given athlete.

2.2.2 Geometric interpretation – hidden clusters

Clustering means finding groupings of the objects such that the objects in
group are more similar to each other than they are to the objects in othe
groups. As we discussed in Chapter 1, this requires a measure of similarit
which is often, in practice, a distance measure of some kind.

Clustering in such a space could be done using any practical measur
of similarity between points, for example Euclidean distance. Unfortunatel
when m is large, such distance measures are not well-behaved. Consider
dataset with m attributes whose values are grouped into three ranges: large
positive, close-to-zero, and large-negative. If attributes are chosen at randon
then the probability of *any* object lying close to the origin is

$$\frac{1}{3^m}$$

Object	Attrib 1	Attrib 2
1	1	2
2	-1	-1
3	2	2
4	0	-1
5	2	1
6	0	-2
7	-1	2

Figure 2.3. *A small dataset.*

because this can happen only if *every* attribute has the value close-to-ze Similarly, if we select an arbitrary object, the probability that any other sin object will be close to it is again $1/3^m$ because the other object's attrib values must match exactly in every position. As m becomes large, th probabilities become smaller and smaller. The geometry of high-dimensio spaces is not intuitive. In a high-dimensional space, objects tend to be from the origin, and every object also tends to be far from every other obje Thus the expected difference between the nearest and furthest neighbor any point is quite small.

There is a natural geometric interpretation of the matrix A, in wh each row defines the coordinates of a point in an m-dimensional space spann by the columns of A. In low-dimensional space, this geometric view can m. it easy to see properties that are difficult to see from the data alone.

For example, suppose we have the small, two-attribute dataset shc in Figure 2.3. A two-dimensional plot of this dataset, as shown in Figure makes it clear that there are two clusters in this data, which is not easy see from the textual form.

Because of their awkward properties, distances in high-dimensional sp. are not as useful for clustering as they might seem.

A matrix decomposition can be interpreted as a transformation of original m-dimensional space into an r-dimensional space. The relationsh (distance, densities) among the objects may be more clearly visible in r-dimensional space that results from the decomposition than they were the geometrical view of the original space. Each row of F is regarded an axis in an r-dimensional space, and the rows of C are the coordinate: each object of A with respect to the space spanned by these axes. Whe well-behaved measure can be defined on this transformed space (which is always possible), clustering can be based on similarity with respect to t measure.

For example, Euclidean distance can be used in r dimensions, where

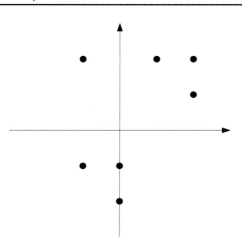

Figure 2.4. *Plot of objects from the small dataset.*

may be a better-behaved measure. In some datasets, the magnitude of the dataset entries is less important than whether the entry is non-zero or not. For example, if the dataset is a *word-document* matrix, with rows corresponding to words, columns corresponding to documents, and the ijth entry counting the frequency of word i in document j, many occurrences of the same word in a document do not necessarily make it more relevant with respect to a search query. Measures based on the angle between the vector from the origin to the point corresponding to each row (or column) may be more appropriate. This angle between two rows can be easily calculated as the *dot product*: the corresponding entries in each row are multiplied together, and the results are summed. The dot product of the two rows is proportional to the cosine of the angle between them, when they are regarded as vectors. The larger the cosine, the smaller the angle between them, and the more similar the two points are.

When the W matrix is diagonal, its entries cause the different dimensions to behave as if they were stretched by different amounts, corresponding to the w_{ii}s. Hence differences in a dimension with a large associated weight are more important than equivalent differences in dimensions with smaller weights. Not all decompositions imply this kind of ranking on the axes – for such techniques we can treat W as the identity matrix.

There is a natural symmetry in this view of a matrix decomposition: the columns of the matrix C can equally well be regarded as defining the axes of an r-dimensional space, and the columns of F as coordinates in this space. Each of these points corresponds to a column of A, that is to one of the attributes of the original dataset.

2.2.3   Component interpretation – underlying processes

Consider the product of the ith column of C, the ith entry of the diago
of W, and the ith row of F. A is the pointwise sum of these *outer prod*
matrices with i ranging from 1 to r.

To see this, let A_i be the matrix obtained by multiplying the ith colu
of C, the ith diagonal element of W and the ith row of F, so

$$A_i = C(:,i) * W(i,i) * F(i,:)$$

A_i has the same shape as A because it is the product of $n \times 1$, 1×1 and $1 \times$
matrices. Now we have that

$$A = \sum_{i=1}^{r} A_i \qquad\qquad (2$$

(to see this, rearrange the sum into the usual form for matrix multiplicati
so each entry in A is the sum of the corresponding entries in each of the
and we can view the A_i as being layers or components whose pointwise s
is the original matrix.

Each layer corresponding to an outer-product matrix can be examin
to see if it can be identified with a known process. For example, one layer n
represent Gaussian noise, or noise with some structural component that ma
it visible via the ordering of the rows of A. Such a layer can be discard
and the remaining layers added back together to give a new version of A fr
which (some of) the noise has been removed. On the other hand, one la
may contain structure that seems fundamental to the modelling task at ha
The matrix corresponding to that layer may be analyzed directly using so
other data-mining technique.

For example, a store may record, each month, the sales of each it
by counting the difference between the number of each item at the beginn
and end of the month, and taking into account any restocking. Howev
most stores suffer losses from shoplifting. Directly analyzing the data me
trying to build a model of two processes at once, sales and shoplifting, an
unlikely to model either accurately. Decomposing the product-stock mat
into two components may produce a component corresponding to sales, an
component corresponding to shoplifting, both of which can then be analy
separately in a principled way.

2.2.4   Graph interpretation – hidden connections

In this interpretation, the entries in the matrix are thought of as strengths
connections between the objects corresponding to the rows and the obje

corresponding to the columns. We can think of A as a bipartite graph (that is with two kinds of nodes or vertices), where one set of nodes corresponds to the objects and the other set of nodes corresponds to the attributes. There are edges between each object-attribute pair (but no edges between objects and objects, or between attributes and attributes). The edge between the node corresponding to object i and the node corresponding to attribute j has a weight associated with it, corresponding to the matrix entry a_{ij}. Of course, if the matrix contains zero entries, we can remove the corresponding edges.

We can consider the edge weights as measures of the strength of the association between object i and attribute j, so that we could imagine that they represent the pull between the nodes at each end. Another useful way to think of these edge weights, however, is as *permeabilities* that represent how easy it is to pass from the node at one end to the node at the other. It now becomes possible to consider how close (that is, how similar) two nodes are in terms of their *commute time*, that is the average 'time' it takes to get from one to the other and back over all possible paths between them. There are connections between such commute distance metrics and the higher-order structure revealed by some decompositions.

A matrix decomposition can be thought of as replacing this bipartite graph by a *tripartite* graph, with three kinds of nodes. The first kind of nodes correspond to the objects, the third kind of nodes correspond to the attributes, but the second kind of nodes correspond to the 'middle' dimension of the matrix decomposition. The number of points in this middle dimension is r. We can think of these middle points as waystations on paths between the points corresponding to objects and attributes. For any given object and attribute there are r different paths, each using a different waystation.

Edge weights can be assigned to each of the edges in the tripartite graph, c_{ij} for the edges from the object nodes to the middle nodes, and f_{jk} for the edges from the middle nodes to the attribute nodes. If a diagonal matrix, W, is part of the decomposition, then its effect can be included by multiplying the weight on each edge by the square root of the corresponding element of the diagonal of W.

Now consider a particular object and attribute, connected in the bipartite version by an edge with weight a_{ik}. In the tripartite graph, this weight, however we interpret it, has been smeared across all of the r paths that join this ith object to the kth attribute via any of the middle-layer points. The relationship between the weights in the tripartite graph and the weight a_{ik} is the standard matrix multiplication relationship

$$a_{ik} = \sum_j c_{ij} f_{jk}$$

This is not an arbitrary decomposition of the value a_{ik} into pieces because the weights c_{ij} and f_{jk} also have to fit into many other such sums for ot object-attribute pairs.

For example, suppose we have a dataset whose rows are people, a whose columns are famous works of art. The people are asked to indica say on a scale of 1 to 10, how much they like any of the works of art w which they are familiar. In the graphical view, these entries become weig on edges linking people to works of art. After the decomposition, these dir links between people and works of art become indirect links passing throu waystations. These waystations may correspond to different groupings of ta in art; perhaps one waystation corresponds to Old Masters and connects sc of the people with paintings by Rembrandt and Rubens. Another waystat may correspond to Impressionists, or those who like Picasso, or Jackson F lock.

The structure of the tripartite graph is reminiscent of a neural netwc especially of an auto-associative neural network.

2.2.5   Summary

Each of the interpretations is simply a way of looking at exactly the sa decomposition – nothing changes in the data, although each interpretat reflects a different view of the meaning of similarity. As humans, howeve different perspective can often provide a new insight into the structures plicit in the data. These different interpretations are different ways of bring our intuitions and understanding of the data to bear. For any particular da set, some of the interpretations may not provide much insight, but it is use to be familiar with them so that they can be used as opportunities arise.

2.2.6   Example

We will make the different interpretations more concrete by considering w light each one sheds on understanding the relationship between two obj in a dataset.

Suppose we are interested in the objects described by rows i and j c dataset. One way to measure their similarity would be to compute their product, the sum of the pointwise products of each row. It is clear intuiti that, if the two objects are similar, then their dot product will be large a positive – because in each position they will tend to have values that are similar magnitudes and sign, and these will contribute to the sum. In fa if the dot product is large and positive, then the rows, as vectors, point similar directions in the obvious m-dimensional geometric space. If the product is zero, then the rows are orthogonal to each other. If the dot prod is negative then, in many positions, the values of the corresponding entries of the rows have opposite signs, and so are very dissimilar. The rows, as vectors point in opposite directions. Dot products are useful ways of formalizing the idea of similarity. If the magnitudes of the values are suitably normalized then dot products correspond to cosines of the angles between the vectors, positive when the angle is small, zero when the angle is a right angle, and negative when the angle is obtuse.

The problem with computing similarity in the space of the original data is that it may misrepresent the real similarities. For example, suppose that the same 'real' attribute was repeated three times in the data (this happens when the attributes appear different but are really just ways of measuring the same thing). Then distances between objects (rows) put three times as much emphasis as they should on the difference between values for the 'real' underlying attribute. This will make a major difference to the apparent similarities, especially for those objects that are far apart according to this 'real' attribute because dot products are squares of Euclidean distance. Adding the square of the distance according to this attribute into the sum three times makes it much larger than if it were only included once. Most matrix decompositions are able to detect this kind of redundancy and provide ways to remove it.

Applying a matrix decomposition makes it possible to compare objects i and j using the ith and jth rows of C rather than A. The *factor* interpretation implies that the entries of a row of C should be thought of as quantities of underlying factors that have been mixed together to produce the observed attribute values of each object. With this view, it might be sensible to look at the *pointwise differences* between the entries of rows i and j of C.

The *geometric* interpretation suggests that we should consider the distances between the points corresponding to objects, in a space of dimension r rather than m and with new axes. This could be done using the Euclidean distance between the points corresponding to rows i and j, or the dot products of rows i and j. Again, a positive dot product indicates similarity, a dot product whose magnitude is small indicates independence, and a negative dot product indicates dissimilarity.

The *component* interpretation suggests that each entry in a row of C comes from a different component. Comparing rows i and j should be done column by column, but the resulting products need not be summed, or combined in any other way. The decomposition may also suggest that some columns should be ignored, and perhaps other columns emphasized.

The *graph* interpretation suggests that the dot product of rows i and j should be interpreted as proportional to the average permeability of the paths between the point corresponding to i and the point corresponding to j. A large value indicates that it is easy to get between these two points on the graph.

We can see from this example that the different interpretations are rea
based on different views of the underlying measures of similarities amo
objects.

2.3   Applying decompositions

The preceding section shows how to understand the results of a matrix
composition in an abstract way. In this section, we consider how mat
decompositions can be applied to extract useful knowledge from a datas
There are two main ways to use matrix decompositions:

1. As discussed earlier, values in many datasets are combinations of d
   from different processes, possibly including noise, that are not part
   the systems that are of interest. Matrix decompositions can be used
   separate the contributions of the different processes, allowing those
   interest to be selected, and those that are not of interest to be discard
   Matrix decompositions can be used for a powerful form of *data clean*

2. Finding the similarities among objects can be difficult because th
   attributes may not all be relevant and because the sheer number
   attributes can make the analysis difficult. Matrix decompositions ma
   it possible to use either standard clustering techniques or specializ
   clustering techniques related to the decomposition more effectively,
   applying them to the C matrix.

Some decompositions can also be used to select interesting objects or
tributes, or to find local relationships within datasets.

2.3.1   Selecting factors, dimensions, components, or waystations

Matrix decompositions break datasets into parts, which may be thought
as underlying factors (the factor interpretation), dimensions (the geome
interpretation), layers (the component interpretation), or waystations (
graph interpretation). This allows us to select one or more parts for remo
or some sets of parts to keep.

Selecting parts of a decomposition is easier if there is some ordering
the parts, for then it is usually clearer how important each part is in explain
the data. Some decompositions naturally impose such an ordering; even wl
they do not, it is still sometimes possible to arrange the parts in a sensi
way. When the entries on the diagonal of W have different magnitudes, tl
these can be sorted into decreasing order, permuting the rows of F and

columns of C to match. When W is the identity matrix, an alternate way to get an ordering is to sort the rows of F in decreasing order of their row norms, permuting the columns of C to match.

When a decomposition is arranged in this way, selection becomes *truncation*, retaining only the *first k* columns of C, the top left $k \times k$ submatrix of W and the first k rows of F. Truncation makes an implicit assumption that the dataset, as given, contains some redundancy or noise whose removal will make it easier to see the 'real' structure in what remains.

This truncated representation is a simplification of the original decomposition. For some matrix decompositions, a bound on how different the new representation is from the old, as a function of k, is possible. The Johnson Lindenstrauss Lemma shows that the number of dimensions in a dataset can be reduced to roughly $\mathcal{O}(\log n)$ without introducing substantial distortion into the L_2 distances between objects (note that this is a function only of n not m).

Denoising

The most obvious reason to remove one or more parts is that we believe those parts to contain noise, either in the sense of noise introduced by measurement or capture of the data, or in the sense of factors that are known to be present but are not of interest (which in the end amounts to the same thing). If noise is present, then some of the parts of the decomposition do not correspond to the process of interest, but to other processes that are somehow artifacts of the data-collection process. Noise removal needs to be carefully considered, however, for two reasons. First, it is hard to be sure that a particular process is truly irrelevant and so safe to remove. Second, the decomposition may not perfectly separate noise processes from processes of interest, so removing the noise may also, unavoidably, remove some of the useful structure.

For example, a dataset of customers who use a store's own credit card might contain information about how much each person has used the card each month, how good their credit rating is, and how many redeemable points they have accumulated. Such a dataset might also contain the number of each credit card, and perhaps the postal code where each customer lives. If we want to cluster the customers to discover if there are useful subgroups of customers who should be treated differently, then we might expect that the credit card numbers are noise that will only obscure the similarities among customers. After all, the credit card numbers are usually handed out sequentially, and similar customers wouldn't be expected to get credit cards at the same time. However, credit card numbers do partly correlate with age, since those who have cards given out a long time ago must be older; and age does correlate with purchasing patterns. So even attributes that seem to have little to contribute

may still be partly useful. Postal codes also have little to do, directly, w
purchasing patterns; but postal codes do correlate with demographics a
wealth, and so also have something indirect to contribute.

If the C, W, and F matrices are truncated at some k ($< r$) then
retain the first k columns of C, the top left hand $k \times k$ submatrix of W (
first k diagonal entries), and the top k rows of F. The product of these sma
matrices

$$C_k W_k F_k$$

is a matrix, A_k, that has the same shape as A. The entries of A_k will dif
slightly from those of A because some of the values that summed to prod
each entry of A are missing (think of the component interpretation). A_k c
be viewed as a version of A from which the effects of some processes h
been removed, leaving an element-wise 'cleaned' version of the dataset.

Of course, the remaining question is how to choose a suitable value
k. There is no definitive answer to this. In practice, what is often done
to consider the values that were used to order the parts (either the values
the diagonal of W or the row norms of F) and look for a sharp change in
values. For example, if these values are plotted, then the plot may show
distinct 'knee' where the values suddenly become smaller. This may sugg
a suitable value for k.

A more sophisticated method, suggested by Zhu and Ghodsi, models
sequences of values used to order the parts explicitly, and assumes that th
values $v_1, \ldots, v_k$ and $v_{k+1}, \ldots, v_m$ are chosen from different distributions.
expression for the profile log-likelihood of k is generated and its maxim
determined exhaustively. This approach seems to work reasonably well
practice. However, this assumes that the truncation is directed at remov
noise, or at least that there are two different major processes from which
data was created.

Another method is based on the fact that multiplying a matrix pointw
by a random $-1, +1$ matrix should change its 2-norm if it contains structu
but will not change it much if the matrix contains no structure. This t
can be applied to the residual matrix, the matrix combining the $k + 1$
m components, to see if it contains any remaining structure that would
quire retaining more components. Again, this method applies only when
truncation is directed at removing noise.

Removing redundancy

The second reason to remove some parts of the data is that the inherent
mensionality of the dataset, that is the number of parts actually needed
lower than the apparent dimensionality in terms of the number of attribu

In other words, the data forms a manifold of k dimensions in a space of n dimensions. This is not attribute selection, but still has the effect of reduc ing the number of attributes for subsequent analysis. Here truncation doe not remove information; rather it expresses it in a more economical form (Of course, as discussed in the previous chapter, real datasets tend to cor tain a number of low-dimensional manifolds oriented in different direction It is still possible, however, that these manifolds together are still of lowe dimension than the entire dataset appears to be.) For example, we prev ously mentioned a dataset of athletes and their heights and weights. If thes are well-correlated, as we expect, then we might see one component in th decomposition representing 'body size', and another component of much le significance, representing variations between height and weight. If this sec ond component is essentially random, then we can remove it, and effectivel compress the two attributes into a single one.

When the diagonal of W is in decreasing order then there is a sensib reason to truncate the C and F matrices after the *first* k rows/column However, it is also possible to select *any* k rows/columns and the matchin diagonal entries of W. The resulting decomposition represents the effec of selecting only certain underlying factors, and enables the interaction different processes to be observed. When the factors are not automaticall ranked, W is the identity matrix, and there is little to guide selection particular submatrices, so many different subsets of k rows/columns may b considered.

Selecting objects and/or attributes with special properties

Selecting parts from a decomposition changes the values in all of the entrie of A but leaves a matrix that is still $n \times m$. However, it is also possibl to use information from the decomposition to remove rows or columns of to produce a new, smaller matrix. Attributes, or objects, or both can b removed.

There are three reasons to remove attributes from the dataset:

1. Some of the attributes are redundant. Recall that many datasets ar collected for other purposes, for example to record transactions. Dat mining of the data is therefore a kind of afterthought. Such datasets ma contain attributes that are not related to the properties of interest. Som prediction and clustering algorithms are not affected by the presence irrelevant attributes, although they always increase the size of the dat and so the time taken to analyze it. However, the majority of data mining algorithms are affected by the presence of unrelated attribute at best spending extra computation time to understand them, and a worst producing lower-quality results because of them. Therefore, i

makes sense to eliminate attributes that have no predictive power or not help to produce better clusterings.

2. Some of the attributes are duplicates. It may happen that several tributes measure essentially the same underlying property, so their v ues are highly correlated. This is a special form of redundancy in wh all but one of each correlated group is redundant. For predictors, presence of such a set of attributes often causes the predictor to spe time understanding the slight (but ultimately uninteresting) differen among the members of the set. For clustering, the presence of such a puts too much emphasis on the dimension corresponding to the und lying property and so skews the clustering. Deciding which membe the set is to be retained is not straightforward – a rule of thumb mi be to keep the best behaved.

3. Some of the attributes are duplicates, but some are much harder collect than the others. In the situation where a set of attributes highly correlated, it might be better to select and retain the attrib whose values are easiest to collect. For subsequent objects, gather the needed data is then much cheaper but the quality of the predict or clustering will stay much the same. Such an attribute is calle *pathfinder* for the other attributes. For example, in medical setti there may be multiple tests that reveal the presence of a particu condition. It is obviously attractive to use the test that is cheapest administer and analyze, or least painful for the patient.

Although selecting attributes sounds like a straightforward process, the re tionships among ordinary attributes, and between an ordinary attribute a the target attribute, are seldom simple. For example, the relationship correlated with" is not usefully transitive in practice, so that two attribu can be predictive of a target attribute, and yet range mutually anywhere fr highly correlated to almost completely uncorrelated.

There are also three reasons to change the objects in a dataset.

1. Some of the objects are outliers. Some clustering techniques do perform well in the presence of outliers, that is single objects that different (far) from all of the other objects. It may be helpful to rem such objects before clustering. For example, if a store is trying to und stand its customers' buying patterns, it may want to ignore custom who have made only a few purchases or purchases worth very little the past year, since there is probably little to learn from them.

2. Some of the objects are almost identical. Replacing a set of alm identical objects by a representative, perhaps with a multiplicity,

reduce the size of the dataset and so the computation time of most data mining techniques. For example, many supermarket customers buy very similar, perhaps identical, sets of products each week.

3. Objects that are the centroids of other sets of objects may be particularly interesting. If the objects are already quite similar, the centroid may represent a 'typical' or 'prototypical' object, to which further analysis can be applied. If the objects seem dissimilar then it is inherently unlikely that their centroid will be an object in the dataset, and this may signal some kind of suspicious or manipulative behavior.

   For example, suppose a dataset contains travel information for a large number of people, and we look for people whose travel patterns are correlated with those of some high-profile person. We would expect such people to perhaps be on the staff of the high-profile person – it's likely that such people have the same basic travel pattern, perhaps sometimes travelling to the same places but earlier. However, *other* people with correlated travel patterns are probably suspicious because they may be conducting surveillance on the high-profile person and should be investigated further [100].

2.3.2   Similarity and clustering

Geometric clustering

We have already discussed how the rows of C can provide a clearer view of the properties of objects than the rows of A. Any data-mining clustering technique can be applied to the data as described by the rows of C and we might expect that the result will be a better clustering than a direct clustering of the data. However, this process assumes that the entries of C can properly be treated as coordinates, and that distances behave as expected. If the axes corresponding to the r rows of F are, for example, not orthogonal, then these assumptions are not correct. This does not mean that clustering will not be effective, but it should be done with some caution and with awareness of F.

Decomposition-based clustering; similarity clustering

All clustering depends on some measure of similarity between objects, or between attributes. We have seen that different interpretations correspond to different views of such measures: the geometric view corresponds to a metric such as Euclidean distance, while the component view corresponds to the elementwise difference.

These interpretations therefore provide either hints or methods for clustering that exploit properties of the decomposition. For example, suppose

that a dataset about customers contains details of their purchasing behavi
but also a customer number. It would be silly to use differences in this c
tomer number as part of a distance measure. (On the other hand, it might *
be sensible to discard it from the dataset, since customer numbers are usua
allocated sequentially in time, and this temporal information often has so
predictive power.)

Graph-based clustering

One particular kind of similarity measure is so different from the others t*
we will discuss it separately. An entire chapter, Chapter 4, will be devoted
it as well.

The similarities among objects in a geometric model are qualitativ
different from the similarities in a graph model. For some datasets, it may
more appropriate to cluster based on a pairwise affinity relationship betw*
objects than to cluster geometrically.

The difference between the two views is the difference between a glo*
view of similarity and a local view of similarity. In a geometric model, *
distance between *any* two objects can be computed, and it stays the sa*
regardless of whether other objects are present or not. On the other hand
a graph model, the distance between any two objects depends on which ot*
objects are present and how they are arranged because the distance depe*
on a path or paths involving all of these objects.

For example, suppose we have a dataset with four attributes. T
objects whose values are $(0, 1, 0, 1)$ and $(1, 0, 1, 0)$ can be directly compa
in a geometric model (their Euclidean distance apart is 2), but they can*
be directly compared in a graph model, and their distance apart depends
other objects that may be present in the dataset. They may not even hav*
well-defined distance between them.

In a larger sense, a geometric space has an existence on its own, *
does not depend on the presence of objects. The shape of a graph space is *
like this at all – all of the distances can be changed by the addition or remc
of a single object. This suggests that care is needed with techniques that
to embed a graph space into a geometric one, for example *multidimensio*
scaling.

2.3.3   Finding local relationships

Although matrix decompositions do not look for local patterns in data
the same way as, say, association rules, they can still be used to look m*
deeply at certain parts of the data. All rows of the dataset matrix are trea

equally by a matrix decomposition but, as it is a numerical technique, it ca be guided by changing the magnitudes of some rows compared to others. Fc example, if the entries in a row of the matrix are multiplied by two, then th will change the decomposition in a way whose effect is to consider that ro as more important than the other rows (twice as important, in fact).

If we know that an object, or for that matter an attribute, is mo important, then this information can be conveyed, indirectly, to the matri decomposition using multiplication of a row or column by a scalar great than one. In the same way, the effect of a row or column can be discounte by multiplying it by a scalar less than one.

This technique can be used to check whether a group of objects or a tributes really are similar to each other, and to decide which of them migl make a good pathfinder. Increasing the importance of one member of th group should have the effect of increasing the importance of the other men bers of the group in a coupled and visible way in the resulting decompositio

This technique can also be used to look for clusters that are completel contained within other clusters. Such hidden clusters may sometimes be d tected directly by density-based clustering but, even when detected, it ma be difficult to find their boundaries. Increasing the importance of one or mo objects suspected to be in the subcluster can have the effect of moving th entire subcluster, relative to the cluster that overlaps it, and so making th subcluster easier to see.

2.3.4 Sparse representations

A matrix is called *sparse* if most of its entries are zero. A decomposition tha results in either C or F being sparse is of interest from the point of view c both analysis and practicality.

If the ith row of the matrix C is sparse, it means that the representatio of object i in the transformed space is a particularly simple combination c the underlying parts. Sparse representations for the objects are attractiv because they increase our confidence that the set of factors captures deepe realities underlying the dataset, and they allow more comprehensible explana tions for the data. For example, some kinds of sparse independent componen analysis seem to correspond to early-stage mammalian vision, where the inpu resources are well understood because they correspond to neurons. Sparse rep resentations are also useful because they reduce the amount of space require to store representations of large datasets.

In the factor interpretation, a sparse row of C means that an object i made up of only a few of the factors. In the geometric interpretation, a spars row means that each object has an extent in only a few dimensions. In th

component interpretation, a sparse row means that each object merges val
from only a few processes. In the graph interpretation, a sparse row mea
that paths from that object to the points corresponding to attributes p
through only a small number of waystations. These statements are all say
the same thing in a different way, but they once again illustrate the power
looking at properties of the data from different perspectives.

The F matrix can also be sparse. When this happens, it suggests t
the parts are themselves particularly simple, requiring only a small amo
of information, and that the parts are quite decoupled from each other.

We have pointed out that any matrix decomposition remains unchang
if the factor matrix is multiplied by an arbitrary invertible matrix, and
coordinate matrix is multiplied by its inverse. This corresponds to a rotat
of the axes of the new space. It is sometimes useful to apply such a rotation
the end, after the decomposition has been computed, with the goal of mak
the representation more sparse. This, of course, reduces the optimality
the solution with respect to whatever criterion was used by the particu
decomposition, but it may nevertheless increase the explanatory power of
result.

2.3.5   Oversampling

We have already mentioned one situation, SemiDiscrete Decomposition, wl
the number of parts, r, of the decomposition is larger than m, the number
attributes. In this case, the decomposition must still generalize the structu
implicit in A because the range of values of the entries of C and F are limit

There is another way in which a decomposition can be 'larger' than
matrix it comes from: when the new parts are redundant. In the decompc
tion, there is now more than one way to describe the same object. Howev
any individual object should not require more than m parts to describe it:
other words, there should not be more than m non-zero entries in any row
C.

Such a decomposition is called an *overcomplete representation*. Th
occur in some natural systems where the parts involved were not globa
optimized (for example, some neural structures in the brain) [80]. They n
also be useful in situations with some inherent ambiguity, for example sig
separation in mobile telephony, where multiple signal paths between a ph
and the local cell tower are commonplace.

2.4 Algorithm issues

The matrices used for data analysis are often very large, so it is useful t
have some sense of the complexity of computing matrix decompositions. Be
cause matrix decompositions are numerical algorithms, it is also important
to be aware of how numerical magnitudes affect results; this can sometime
cause computational problems such as instability, but can also be exploite
to discover finer details of the structure present in the data.

2.4.1 Algorithms and complexity

Even looking at all of the elements of A has complexity $\Theta(nm)$, and it is har
to see how a useful matrix decomposition could avoid complexity $\Omega(nmr$
In practice, most matrix decompositions are much more expensive, perhap
quadratic in one or both of n and m. Although quadratic complexity does no
sound alarming, n can be extremely large so the execution time to comput
a matrix decomposition is often a limitation in practice.

Because n is often so large in real-world applications, matrix decompo
sitions may not even be compute-bound. The performance bottleneck ma
actually be the time required to fetch the entries of A from the bottom o
the memory hierarchy. This requires $\Theta(nm)$ operations, but the constant
required in modern architectures are very large, typically comparable in siz
to m. Hence memory access times can be as bad as computation times.

For these reasons, there is a great deal of ongoing research aimed a
exploiting sparse matrix algorithms; computing approximate matrix decom
positions, for example low-rank approximations; and exploiting quantizatio
of the matrix entries.

2.4.2 Data preparation issues

Many datasets have attributes that are categorical, that is they have value
for which no natural ordering exists. For example, customers may pay usin
a range of named credit cards. Since matrix decompositions are numerica
techniques, these categorical attributes must be converted into numeric values
Care must be taken not to introduce spurious correlations because of th
conversion. Techniques require a tradeoff between the expense of additiona
attributes and the accuracy of the mapping. The best approach is to ma
each categorical value to a corner of a generalized tetrahedron, requiring n -
1 new attributes for n categorical values. A cheaper approach is to ma
the categorical values to equally spaced points on a circle (requires 2 ne
attributes) or sphere (requires 3 new attributes). Of course, the order of th
mapping around the circle or sphere must still be given some consideration.

2.4.3 Updating a decomposition

In many situations, data continues to become available after an initial mat
decomposition has been computed. Such data can be considered as new ro
for the matrix A. There are several ways to include this new data in
existing decomposition, that is create new rows of the matrix C correspond
to the new data, and adjusting W and F to include the new informat
implicit in the new data.

The first and simplest way is to add the new rows to A and repeat
decomposition. This fully incorporates the information from the new data
the model, but it is an expensive solution because of the high complexity
the algorithms for computing decompositions.

The second way is to use an incremental algorithm that includes
new information without carrying out a new decomposition from scratch.
most matrix decompositions, such incremental algorithms are known. Th
complexity is usually much less than the complexity of a new decompositi

If the matrices F and W are invertible, then $F^{-1}W^{-1}$ can be viewed
transforming data that looks like rows of A into data that looks like rows
C. Applying this transformation approximates the effect of decomposing
larger matrix that combines A and the new data into a new larger C –
note that neither W nor F is changed, so the new data does not change
model. This transformation only allows us to see what the new data wo
look like in the context of the original decomposition.

This can nevertheless be quite useful. If the original matrix captu
enough data about the problem domain, then its decomposition reveals
implicit structure of this domain. If this implicit structure does not cha
with time, then applying the transformation to new data shows the underly
properties of the new data, even though the structure is not being upda
to reflect the new data.

For example, if the original decomposition led to a clustering of
rows of A, multiplying new objects by $F^{-1}W^{-1}$ maps them to locations
the space of the clustering. Such objects can then be allocated to the clust
to which they are closest.

Although the way in which C and F are *combined* to give A is line
it is perfectly possible to construct C and F in some other, non-linear, w
We will not discuss this further, but it shows that matrix decompositions
decompose data in even more sophisticated ways.

Notes

Hubert *et al.* provide a good historical comparison of the role of matrix decompositions, contrasting their use in linear algebra and in data analysis [57].

The standard references for the matrix decompositions we will use are Singular value decomposition (SVD) [48]; SemiDiscrete Decomposition (SDD) [72, 73]; Independent Component Analysis (ICA) [58, 60]; Non-Negative Matrix Factorization (NNMF) [78]; but see the relevant chapters for a fuller list of references.

Chapter 3

Singular Value Decomposition (SVD)

The *singular value decomposition* (SVD) transforms the data matrix in a way that exposes the amount of variation in the data relative to a set of later features. The most natural interpretation is geometric: given a set of data in m-dimensional space, transform it to a new geometric space in which as much variation as possible is expressed along a new axis, as much variation independent of that is expressed along an axis orthogonal to the first, and so on. In particular, if the data is not inherently m-dimensional, its actual dimensionality (the rank of the data matrix, A) is also exposed.

3.1   Definition

The singular value decomposition of a matrix A with n rows and m columns is

$$A = USV'$$

where the superscript dash indicates the transpose of matrix V.

If A has rank r, that is r columns of A are linearly independent, then U is $n \times r$, S is an $r \times r$ diagonal matrix with non-negative, non-increasing entries $\sigma_1, \sigma_2, \ldots, \sigma_r$ (the singular values), and V' is $r \times m$. In addition, both U and V are orthogonal, so that $U'U = I$ and $V'V = I$. This is actually the so-called 'thin' SVD. If all of the singular values are different, the SVD is unique up to multiplication of a column of U and the matching row of V' by -1. In most practical datasets, $r = m$, since even if several attributes (that is columns) are really measurements of the same thing, which is the commonest way in which the rank of A would be less than m, they are typically not *exactly* correlated.

By convention, the third matrix in the decomposition is written a transpose. This emphasizes the duality between objects and attributes cause both U and V are matrices whose rows correspond to objects a attributes respectively, and whose columns correspond to the r new pa Unfortunately, this makes it easy to make mistakes about which way rov V is considered, and which are its rows and columns.

The natural interpretation for an SVD is geometric (Section 3.2.2), the component interpretation is also useful (Section 3.3.1).

Recall our example dataset matrix, introduced on Page 5. The U, S and matrices of the singular value decomposition of this matrix are:

$$U = \begin{bmatrix}
-0.31 & -0.40 & 0.29 & 0.25 & 0.13 & -0.07 & -0.27 & 0.07 \\
-0.37 & -0.31 & 0.18 & 0.10 & 0.30 & 0.28 & -0.13 & -0.38 \\
-0.31 & -0.23 & -0.63 & -0.03 & -0.38 & 0.23 & -0.23 & -0.29 \\
-0.37 & 0.48 & -0.31 & -0.20 & 0.18 & 0.29 & 0.20 & 0.04 \\
-0.33 & 0.55 & 0.45 & 0.03 & -0.02 & 0.17 & -0.23 & -0.13 \\
-0.22 & -0.23 & -0.19 & -0.02 & 0.53 & -0.02 & 0.31 & 0.38 \\
-0.24 & 0.03 & -0.03 & 0.00 & -0.15 & 0.14 & -0.35 & 0.74 \\
-0.19 & 0.07 & -0.02 & 0.23 & 0.16 & -0.07 & 0.47 & -0.16 \\
-0.29 & 0.19 & -0.21 & -0.06 & 0.15 & -0.83 & -0.29 & -0.12 \\
-0.30 & 0.02 & 0.06 & 0.56 & -0.51 & -0.16 & 0.39 & 0.12 \\
-0.32 & -0.25 & 0.32 & -0.72 & -0.32 & -0.14 & 0.30 & 0.01
\end{bmatrix}$$

$$S = \begin{bmatrix}
41.50 & 0.00 & 0.00 & 0.00 & 0.00 & 0.00 & 0.00 & 0.00 \\
0.00 & 13.22 & 0.00 & 0.00 & 0.00 & 0.00 & 0.00 & 0.00 \\
0.00 & 0.00 & 7.92 & 0.00 & 0.00 & 0.00 & 0.00 & 0.00 \\
0.00 & 0.00 & 0.00 & 7.57 & 0.00 & 0.00 & 0.00 & 0.00 \\
0.00 & 0.00 & 0.00 & 0.00 & 4.15 & 0.00 & 0.00 & 0.00 \\
0.00 & 0.00 & 0.00 & 0.00 & 0.00 & 3.47 & 0.00 & 0.00 \\
0.00 & 0.00 & 0.00 & 0.00 & 0.00 & 0.00 & 2.37 & 0.00 \\
0.00 & 0.00 & 0.00 & 0.00 & 0.00 & 0.00 & 0.00 & 0.52
\end{bmatrix}$$

$$V = \begin{bmatrix}
-0.29 & 0.63 & 0.04 & -0.03 & 0.65 & 0.20 & -0.25 & -0.02 \\
-0.37 & 0.09 & -0.53 & -0.45 & -0.36 & 0.20 & -0.09 & -0.46 \\
-0.36 & 0.33 & 0.29 & 0.04 & -0.19 & -0.03 & 0.80 & -0.05 \\
-0.35 & -0.05 & -0.44 & 0.16 & -0.10 & 0.20 & 0.04 & 0.78 \\
-0.38 & 0.08 & 0.38 & -0.34 & -0.25 & -0.56 & -0.38 & 0.25 \\
-0.36 & -0.20 & -0.35 & 0.47 & 0.27 & -0.58 & 0.05 & -0.28 \\
-0.36 & -0.11 & 0.36 & 0.55 & -0.31 & 0.41 & -0.34 & -0.20 \\
-0.36 & -0.65 & 0.23 & -0.37 & 0.42 & 0.23 & 0.16 & -0.02
\end{bmatrix}$$

SVD and Principal Component Analysis (PCA)

Although this is not the way we will think about SVD in this chapter, SVD is intimately connected with eigenvectors and eigenvalues. Principal component analysis (PCA) is another way to understand data, and there is considerabl disagreement about the differences between the two techniques. Some author consider them to be identical, others to differ in normalization strategies, an still others consider them to be completely distinct.

Most versions of principal component analysis find the eigenvectors an eigenvectors of either the matrix AA', which describes the correlation amon the objects, or the matrix $A'A$, which describes the correlation among th attributes. Much of what is said about SVD in this chapter also holds fc PCA, but PCA is limited in at least the following two ways: first, it analyze either the objects or the attributes independently, whereas SVD analyzes bot together; and second, the correlation matrices are expensive to form (AA' i $n \times n$ which makes it difficult to handle) and often ill-conditioned, so tha computing the eigenvectors is problematic.

Normalization

Because SVD is a numerical algorithm, it is important to ensure that th magnitudes of the entries in the dataset matrix are appropriate, so that prop erties are compared in a way that accords with comparisons in the real worlc For example, height and weight are roughly correlated in humans. Howevei if height is measured in miles, and weight in grams, then weight is going t seem much more important during the decomposition.

In general we don't know what the 'right' units are for each attribute. I the absence of better information, the only sensible thing to do is to scale all c the attribute values into roughly the same range. This encodes an assumptio that all attributes are of about the same importance. This is quite a stron assumption, but it is hard to see how to do better.

If the values in the data matrix A are all positive (say), the first compc nent of the decomposition will capture the rather trivial variation along th axis that joins the origin to the centroid of the data (in m-dimensional space We could, of course, ignore this component in subsequent analysis. The prot lem is, however, that the new axes are forced to be orthogonal to each othei so that the second axis points in a distorted direction. This is illustrate in Figure 3.1, where the top ellipse shows what happens when positive dat is transformed. The second axis does not properly capture variation in th data because of the existence of the first axis. The bottom ellipse shows wha happens when the data is *zero centered*, that is, for each column, the mea of that column is subtracted from each entry. This moves the data 'clouc

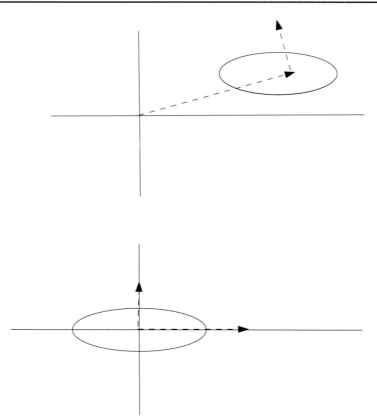

Figure 3.1. *The first two new axes when the data values are posi* *(top) and zero-centered (bottom).*

so that it is centered at the origin. Now the new axes correctly capture directions of variation in the data.

To address the possibly different magnitudes of different attributes, i usual to divide the entries in each column by the standard deviation of t column. This has the effect of scaling the values in each column so that m of them fall in the range -1 to $+1$ – but also wires in the assumption t the distribution of values in each column is approximately Gaussian. By Law of Large Numbers, this is a plausible assumption, but its existence sho always be remembered. In particular, if several attributes have distributi that are far from Gaussian, Independent Component Analysis may be a bet decomposition (see Chapter 7). Values that have been transformed by z centering and division by the standard deviation are known as *z scores*.

When the significance of magnitudes is non-linear, for example wl

very large values are present in the data but do not have corresponding
large significance, the attribute values can be scaled by taking logarithms (
the values are all positive) or by taking cube roots (if they are both positiv
and negative), and then transforming to z scores.

Our example matrix, normalized using z-scores, is:

$$
A = \begin{bmatrix}
-0.97 & -1.09 & -0.53 & -0.33 & 0.38 & 0.89 & 1.40 & 1.30 \\
-0.32 & -0.20 & -0.053 & 0.42 & 0.38 & 0.89 & 1.40 & 1.64 \\
-0.97 & 1.59 & -1.01 & 1.92 & -0.77 & 0.89 & -0.11 & 0.27 \\
1.62 & 1.59 & 1.38 & 1.17 & 0.38 & -0.44 & -0.62 & -0.76 \\
1.62 & -0.20 & 1.86 & -1.08 & 1.54 & -1.78 & 0.90 & -1.11 \\
-0.65 & -0.64 & -1.01 & -0.33 & -1.35 & 0.22 & -1.12 & 0.61 \\
-0.32 & -0.20 & -0.53 & -0.33 & -0.19 & -1.11 & -0.11 & -0.42 \\
-0.32 & -1.09 & -0.05 & -1.08 & -1.35 & -0.44 & -0.62 & -0.76 \\
0.32 & 0.25 & -0.05 & -0.33 & 0.96 & 0.89 & -1.12 & -0.76
\end{bmatrix}
$$

The resulting U, S and V matrices are:

$$
U = \begin{bmatrix}
-0.30 & 0.35 & -0.30 & -0.15 & -0.15 & 0.10 & -0.21 & 0.13 \\
-0.21 & 0.08 & -0.47 & -0.10 & -0.06 & 0.46 & 0.00 & -0.21 \\
-0.37 & -0.57 & -0.12 & 0.15 & 0.27 & -0.25 & -0.21 & -0.40 \\
0.39 & -0.58 & -0.04 & -0.00 & 0.03 & 0.35 & 0.26 & 0.28 \\
0.66 & 0.19 & -0.07 & -0.23 & 0.13 & 0.09 & -0.29 & -0.26 \\
-0.31 & 0.07 & 0.42 & 0.13 & -0.04 & 0.40 & 0.18 & 0.29 \\
0.00 & 0.08 & 0.24 & 0.08 & 0.37 & -0.15 & -0.56 & 0.45 \\
-0.04 & 0.21 & 0.56 & -0.16 & 0.14 & 0.02 & 0.27 & -0.51 \\
0.08 & -0.15 & 0.18 & 0.07 & -0.84 & -0.25 & -0.20 & -0.05 \\
-0.06 & 0.01 & -0.16 & -0.53 & 0.08 & -0.52 & 0.44 & 0.28 \\
0.18 & 0.30 & -0.24 & 0.74 & 0.08 & -0.25 & 0.33 & -0.00
\end{bmatrix}
$$

$$
S = \begin{bmatrix}
5.35 & 0.00 & 0.00 & 0.00 & 0.00 & 0.00 & 0.00 & 0.00 \\
0.00 & 4.26 & 0.00 & 0.00 & 0.00 & 0.00 & 0.00 & 0.00 \\
0.00 & 0.00 & 3.69 & 0.00 & 0.00 & 0.00 & 0.00 & 0.00 \\
0.00 & 0.00 & 0.00 & 3.57 & 0.00 & 0.00 & 0.00 & 0.00 \\
0.00 & 0.00 & 0.00 & 0.00 & 1.86 & 0.00 & 0.00 & 0.00 \\
0.00 & 0.00 & 0.00 & 0.00 & 0.00 & 1.46 & 0.00 & 0.00 \\
0.00 & 0.00 & 0.00 & 0.00 & 0.00 & 0.00 & 1.08 & 0.00 \\
0.00 & 0.00 & 0.00 & 0.00 & 0.00 & 0.00 & 0.00 & 0.27
\end{bmatrix}
$$

$$V = \begin{bmatrix} 0.50 & -0.23 & 0.07 & -0.22 & -0.21 & 0.70 & -0.33 & -0.03 \\ 0.14 & -0.59 & -0.22 & 0.42 & 0.13 & -0.14 & -0.04 & -0.60 \\ 0.53 & -0.01 & -0.21 & -0.25 & 0.04 & -0.05 & 0.78 & 0.00 \\ -0.20 & -0.64 & -0.28 & -0.12 & 0.23 & 0.02 & 0.01 & 0.64 \\ 0.37 & 0.17 & -0.49 & 0.31 & -0.47 & -0.33 & -0.28 & 0.31 \\ -0.43 & -0.22 & -0.16 & -0.36 & -0.73 & -0.02 & 0.16 & -0.23 \\ -0.05 & 0.23 & -0.60 & -0.53 & 0.37 & -0.05 & -0.31 & -0.27 \\ -0.31 & 0.24 & -0.45 & 0.44 & 0.02 & 0.61 & 0.27 & 0.00 \end{bmatrix}$$

If we compare the two sets of matrices resulting from the SVD of unnormalized and normalized versions of A, we see a large difference in singular values. For the unnormalized version, the largest singular value 41.50, followed by 13.22 and 7.92 For the normalized version, the larg singular value is 5.35, followed by 4.26 and 3.69. The large singular va for the unnormalized matrix reflects the average value of the matrix ent or the length of the dashed vector in the situation at the top of Figure 3.1

When a matrix is sparse, that is most of its entries are zeros, it may more appropriate to normalize by keeping the zero entries fixed. The m of the non-zero entries is then subtracted from the non-zero entries so t they become zero-centered, and only the non-zero entries are divided by standard deviation of the column mean. This form of normalization need be considered carefully because it reduces the impact of zero values on way other values are adjusted, and so should not be used if zero values h some special significance. There is also an issue of how many non-zero ent there should be in the matrix before it is no longer considered sparse.

3.2 Interpreting an SVD

Although, as we have said, the geometric interpretation is most natural an SVD, there is something to be learned from the other interpretations.

3.2.1 Factor interpretation

Interpreting the rows of V' (the columns of V) as underlying factors is perh the oldest way of understanding an SVD. For example, suppose we want understand what makes people happy. We might suspect that factors such income, education, family life, marital status, and a satisfying job might all relevant, but we couldn't be sure that these were all the factors, and we mi not be sure precisely how to measure them. Designing a questionnaire to about such factors, and also about degree of happiness, might need questi directly about income, but also questions about home ownership, pens

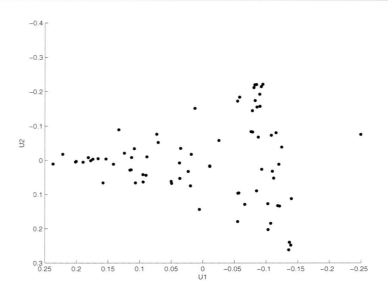

Figure 3.2. *The first two factors for a dataset ranking wines.*

plan, and medical insurance. It might turn out that all of these correlat
strongly with income, but it might not, and the differences in correlatio
may provide insight into the contribution of a more general concept such a
'prosperity' to happiness. The survey data can be put into a matrix wit
one row for each respondent, and one column for the response each question
An SVD of this matrix can help to find the latent factors behind the explici
factors that each question and response is addressing.

For datasets of modest size, where the attributes exhibit strong correla
tions, this can work well. For example, Figure 3.2 is derived from a dataset i
which 78 people were asked to rank 14 wines, from 1 to 14, although many di
not carry out a strict ranking. So the attributes in this dataset are wines, an
the entries are indications of how much each wine was liked by each person
The figure shows a plot along the first two axes of the transformed space
corresponding to the two most important factors. Some further analysis i
required, but the first (most important) factor turns out to be liking for win
– those respondents at the left end of the plot are those who like wine, tha
is who had many low numbers in their 'ranking', while those at the right en
liked wine less across the board. This factor corresponds to something whic
could have been seen in the data relatively easily since it correlates strongl
with the sum of the 'rankings'. For example, the outlier at the right en
corresponds to someone who rated every wine 14.

The second factor turns out to indicate preference for red versus whit
wine – those respondents at the top of the plot prefer red wine over white

while those at the bottom of the plot prefer white over red. This factor is m
less easy to see directly in the data. Notice also that the matrix decomposit
does not know the 'meaning' of any column of the dataset. It discovers t
pattern by noticing that certain rankings are correlated with other rankin

One obvious conclusion that can be drawn just from seeing the triangu
shape of the plot in the figure is that those who like wine a lot do not h
strong preferences for red versus white; it is those who like wine less who t
to have such a preference. These simple results have immediate implicati
for marketing wine.

However, for large complex datasets, the factors tend to be linear con
nations of all or most of the attributes in the dataset because each attribut
partially correlated with many of the others in subtle ways. Hence, it is of
difficult to interpret the factors and relate them to the application doma
from which the original attributes come.

3.2.2   Geometric interpretation

The geometric interpretation of an SVD is to regard the rows of V (colun
of V') as defining new axes, the rows of U as coordinates of the objects
the space spanned by these new axes, and S as a scaling factor indicat
the relative importance (or stretching) of each new axis. Note that the p
sible non-uniqueness of the decomposition means that an axis can be flip
without changing anything fundamental.

Because the SVD is symmetric with respect to rows and columns
can also be regarded as defining a new space spanned by the rows of U a
mapping the attributes from coordinates in an original n-dimensional sp
into this new space. The maximum variation among the attributes is captu
in the first dimension, and so on.

The most useful property of the SVD is that the axes in the new spa
which represent new pseudoattributes, are orthogonal. Hence the expl
properties of each object as characterized by the original attributes are
pressed in terms of new attributes that are independent of each other. As
saw in Figure 3.2, the orthogonality of the new axes means that the rows
the C matrix can be plotted in space in a way that accurately reflects tl
relationships.

Rotation and stretching

There are several intuitive ways to understand how an SVD is transform
the original data.

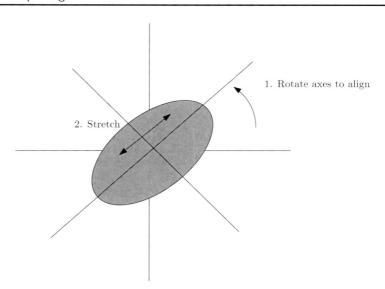

Figure 3.3. *One intuition about SVD: rotating and scaling the axes.*

First, notice that when we interpret the rows of A as coordinates in a m-dimensional space, the axes of this space can be made explicit by writing A as AI, where I is the m-dimensional identity matrix. These axes are just the ordinary Cartesian axes. The matrix decomposition is asserting the equivalence of these coordinates and ordinary axes to new coordinates (the rows of U) and a new set of axes described by the product SV'. The matrix V' is a rotation, and the matrix S is a stretching of the axes. However, these axes are not arbitrary; rather they have been computed based on the data itself.

Imagine the unit sphere in m dimensions. Then V' followed by S rotates and stretches this unit sphere so that it fits 'over' the data. This fitting guarantees that the coordinates required to describe each object will be as simple as possible. Figure 3.3 illustrates the process. The gray ellipse represents the raw data. First the axes are rotated to align with the axes of the rough ellipse formed by the data. Then the axes are stretched so that they better fit the extents of the data. Relative to these new axes, the coordinates of each data point are simpler.

This intuition also shows clearly the effect of normalization on the SVD. Zero centering places the rough ellipse corresponding to the data close to the origin. Dividing the entries in each column by their standard deviation from the mean makes the structure of the data as close to a sphere as possible – so that the rotation and scaling can concentrate on the *distribution* or *density* of the objects in each direction.

A special case that can also be understood from this intuitive point of

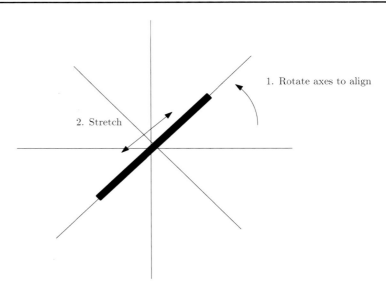

Figure 3.4. *Data appears 2-dimensional but can be seen to be dimensional after rotation.*

view is when the raw data appears to have dimensionality m but is actuall lower-dimensional manifold. This situation is shown in Figure 3.4. Here raw data seems to have dimension two – each data point requires an x an coordinate to describe its position (or the value of two attributes). Howev the rotation shows that only one new axis and one stretch factor are requi to fully describe the dataset.

Springs

.

     Another helpful way to think about the transformation that SVD doe the following. Suppose we place points corresponding to both the objects the attributes in the same m-dimensional space. Now connect the ith obj to the jth attribute by a spring whose tension corresponds to the magnit of the ijth entry of the matrix A. (If the entry is negative, then the sprin repulsive rather than attractive.) Then the stable positions where each pc is at rest correspond to the locations described by the U matrix, for obje and the V matrix, for attributes. Actually, to get the scaling right, th locations correspond to locations described by $US^{1/2}$ and $VS^{1/2}$ because singular values describe the scaling of each dimension relative to the othe

     This view of SVD illustrates the symmetry between objects and tributes. It also shows how SVD makes use of the indirect and higher-or relationships among object and attributes. If we assume that all entries of the matrix are non-zero, then the relative position of two objects depends on the positions of all of the attributes; but these, in turn, depend on the position of these two objects, but also of all of the other objects. The position of these objects depends on the strength of their connection to the attributes, and so on. The SVD represents the fixed point of this reasoning (the stable position into which the points settle) and so takes into account *all* of the correlation information between objects and attributes.

Dot products

Another way to understand the geometry of an SVD is this: if each point in m-dimensional space is regarded as a vector, that is we associate each point with the line from the origin to that point, then the angles between vectors represent the correlation between them. The cosine of the angle between them is the dot product of the two vectors, with their lengths appropriately scaled

$$\text{cosine of angle between } x \text{ and } y = \frac{<x, y>}{<x, x><y, y>}$$

where $< x, y >$ indicates the dot product of two vectors. Hence two vectors in roughly the same direction represent two correlated objects, and their dot product will be a large positive number. Two vectors that point in opposite directions are negatively- or anti-correlated and their dot product will be a large negative number. Two vectors that are orthogonal to each other are uncorrelated and will have dot product 0.

Consider a vector that is *uncorrelated* with many of the other vectors. Its position will have to be such that its dot product is (close to) zero with all of these other vectors. Unfortunately, there are only m different, orthogonal directions to point, and n is much larger than m. The only other way to have a small dot product with many other vectors is to be a short vector, whose endpoint plots close to the origin. Hence, objects that are largely uncorrelated with other objects tend to be plotted close to the origin. This property holds even in spaces of lower dimension. The vector components in the higher dimensions have very little effect on the magnitude of dot products because the corresponding singular values are so small. Hence, taking dot products using the first k components of vectors produces dot products whose magnitudes are close to their 'true' magnitudes, and so approximate the correlation structure.

Now consider a vector that is correlated with *many* of the other vectors. Its position will have to be such that its dot product with many other vectors will be large; in other words, it wants to point in many directions. The net result, again, is that such vectors will be small, and the corresponding objects are plotted close to the origin.

Objects that are plotted far from the origin tend to be those wh correlation with the other objects is intermediate; neither highly-correla nor uncorrelated. Hence an SVD has a built-in filter for interestingness. I transformed and truncated space, objects that are either correlated with other objects, or with almost all other objects tend to be short vectors, end close to the origin. On the other hand, objects with unusual correlation w the other objects tend to be long vectors, ending far from the origin.

This is useful in two related ways. First, the values of the new attribu (the columns of V') provide some information about the importance of objects – objects that are far from the origin along some dimension are usua significant. For example, recall the first dimension of the wine dataset, wh could be used to identify those who liked wine most and least. However, sc ing the objects according to their distance from the origin in the transform space, using several attributes, is a much more significant ranking, since places both objects that are 'important' only in some commonplace way, a objects that are idiosyncratic, at the bottom of the list. The objects at top of the ranking (those that are furthest from the origin in the transform space) are those that are interesting in a far more subtle way.

Second, objects that are close to the origin, say closer than the med distance, are the least interesting and can be discarded from further analy reducing the size of the dataset in a way guaranteed to preserve the m interesting structure.

3.2.3   Component interpretation

Let u_i be the ith column of U, s_i the ith singular value of S and v_i the row of V. Then

$$A = \sum_{i=1}^{m} A_i$$

where $A_i = u_i s_i v'_i$. This sum tells us that we can think of each entry o as the sum of the corresponding entries in each of the A_i, and of A as pointwise sum of the A_i. In other words, the A_i form layers that toget recreate A.

This view is exactly what we hypothesised was true for many real-wo datasets: the value of a particular entry in the dataset is the result of superposition of a number of processes, only some of which are of inter For SVD, the layers represent independently varying values.

Of course, there is no necessary reason why the decomposition into lay that an SVD provides should correspond to the set of underlying proces that were at work when the dataset was collected, but a correspondence often be found in practice, at least in the earlier dimensions.

For our example matrix, A, the matrices A_1 and A_2 are:

$$
A_1 = \begin{bmatrix}
-0.80 & -0.23 & -0.85 & 0.33 & -0.60 & 0.69 & 0.07 & 0.50 \\
-0.57 & -0.16 & -0.60 & 0.23 & -0.42 & 0.49 & 0.05 & 0.35 \\
-0.99 & -0.28 & -1.05 & 0.41 & -0.74 & 0.85 & 0.09 & 0.62 \\
1.04 & 0.29 & 1.10 & -0.43 & 0.77 & -0.89 & -0.10 & -0.65 \\
1.74 & 0.49 & 1.85 & -0.72 & 1.30 & -1.50 & -0.16 & -1.09 \\
-0.82 & -0.23 & -0.87 & 0.34 & -0.62 & 0.71 & 0.08 & 0.51 \\
0.00 & 0.00 & 0.00 & -0.00 & 0.00 & -0.00 & -0.00 & -0.00 \\
-0.11 & -0.03 & -0.12 & 0.05 & -0.08 & 0.10 & 0.01 & 0.07 \\
0.21 & 0.06 & 0.22 & -0.09 & 0.16 & -0.18 & -0.02 & -0.13 \\
-0.17 & -0.05 & -0.18 & 0.07 & -0.12 & 0.14 & 0.02 & 0.10 \\
0.47 & 0.13 & 0.50 & -0.19 & 0.35 & -0.40 & -0.04 & -0.29
\end{bmatrix}
$$

$$
A_2 = \begin{bmatrix}
-0.34 & -0.88 & -0.01 & -0.95 & 0.25 & -0.33 & 0.34 & 0.36 \\
-0.08 & -0.21 & -0.00 & -0.22 & 0.06 & -0.08 & 0.08 & 0.09 \\
0.56 & 1.45 & 0.02 & 1.55 & -0.40 & 0.54 & -0.55 & -0.60 \\
0.57 & 1.48 & 0.02 & 1.58 & -0.41 & 0.55 & -0.56 & -0.61 \\
-0.19 & -0.49 & -0.01 & -0.53 & 0.14 & -0.18 & 0.19 & 0.20 \\
-0.07 & -0.17 & -0.00 & -0.18 & 0.05 & -0.06 & 0.06 & 0.07 \\
-0.08 & -0.21 & -0.00 & -0.23 & 0.06 & -0.08 & 0.08 & 0.09 \\
-0.21 & -0.54 & -0.01 & -0.58 & 0.15 & -0.20 & 0.21 & 0.22 \\
0.14 & 0.37 & 0.01 & 0.40 & -0.10 & 0.14 & -0.14 & -0.15 \\
-0.01 & -0.03 & -0.00 & -0.03 & 0.01 & -0.01 & 0.01 & 0.01 \\
-0.29 & -0.76 & -0.01 & -0.82 & 0.21 & -0.29 & 0.29 & 0.31
\end{bmatrix}
$$

and their sum is

$$
\begin{bmatrix}
-1.14 & -1.11 & -0.87 & -0.61 & -0.36 & 0.36 & 0.41 & 0.87 \\
-0.65 & -0.37 & -0.60 & 0.01 & -0.37 & 0.41 & 0.13 & 0.44 \\
-0.43 & 1.17 & -1.03 & 1.95 & -1.14 & 1.39 & -0.46 & 0.02 \\
1.60 & 1.77 & 1.12 & 1.16 & 0.36 & -0.34 & -0.66 & -1.25 \\
1.55 & -0.00 & 1.84 & -1.24 & 1.44 & -1.68 & 0.03 & -0.88 \\
-0.89 & -0.40 & -0.88 & 0.16 & -0.57 & 0.64 & 0.14 & 0.58 \\
-0.08 & -0.21 & -0.00 & -0.23 & 0.06 & -0.08 & 0.08 & 0.09 \\
-0.32 & -0.57 & -0.13 & -0.53 & 0.07 & -0.11 & 0.22 & 0.29 \\
0.35 & 0.43 & 0.23 & 0.31 & 0.05 & -0.04 & -0.16 & -0.29 \\
-0.18 & -0.08 & -0.18 & 0.03 & -0.11 & 0.13 & 0.03 & 0.12 \\
0.18 & -0.63 & 0.49 & -1.01 & 0.56 & -0.69 & 0.25 & 0.02
\end{bmatrix}
$$

3.2.4 Graph interpretation

The graph interpretation of SVD takes a bipartite graph, whose two kinds of objects correspond to objects and to attributes, and whose edges are weighted by the entries of the matrix, and expands it to a tripartite graph. In this partite graph, there is a third set of r waystation vertices corresponding the 'middle' dimension of the SVD. The vertices corresponding to the obje are fully connected to the waystation vertices that are created by the decc position; and these in turn are fully connected to the vertices correspond to the attributes.

Each edge in the tripartite graph has an associated weight. Those c necting objects to waystations get their weights from the entries of the mat $US^{1/2}$, and those connecting waystations to attributes get their weights fr the entries of the matrix $VS^{1/2}$. The fact that the product of these matri is A means that these weights fit together properly. The sum of the weig along all of the paths between a particular object i and an attribute j the ijth entry of A, as long as the weights *along* a path are accumulated multiplication. These weights can be understood as capturing the simila between the vertices they connect; or equivalently the permeability of connection between them, or how easy it is to travel from one end to other.

The intuition here is that an SVD allocates capacity to each edge optimize the total permeability of all paths. The weight associated with edge from, say, an object to a waystation must be assigned so that it fits w the paths from that object to *all* of the attributes, since this path make contribution to all of them.

3.3 Applying SVD

3.3.1 Selecting factors, dimensions, components, and waystations

The main distinguishing feature of an SVD is that it concentrates variat into early dimensions. This means that the natural way to select parts of structure inside the dataset is to select, from the r components, the first

We have suggested that there are two main reasons to select and ret only some parts of a decomposition: because the discarded parts are cons ered noise; or because the discarded parts represent some process that we not wish to model. Given the ordering of the parts by an SVD, these decisi are much the same. The only difference is that we might use slightly differ criteria to choose how many parts to retain and how many to discard.

Suppose that we want to represent the dataset properties in a space dimension k (where $k \leq r$), that is we want to retain only k parts of decomposition. The first k rows of V' define the axes of a k-dimensio space. Surprisingly good representations of spaces with many hundreds dimensions can be achieved by quite small values of k, perhaps less than

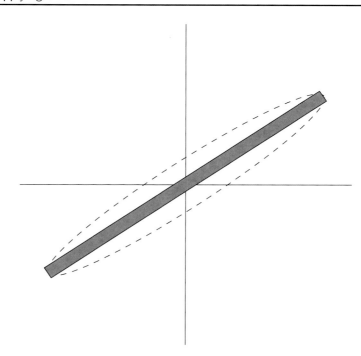

Figure 3.5. *The effect of noise on the dimensionality of a dataset.*

Denoising

A dataset that contains noise may appear to be of much higher dimensionalit
than it really is. Figure 3.5 shows, as a dark line, a 1-dimensional datase
with two (perfectly correlated) attributes, and so appearing 2-dimensiona
As discussed above, an SVD will quickly detect that the data is actually 1
dimensional. The dashed ellipse shows what happens when noise is added t
the dataset. The data now has an apparent extent in the second direction.

After the SVD transformation the data will appear to be 2-dimensiona
– but the extent, and so the amount of stretching, required in the secon
dimension will be small. This is the clue that this dimension does not contai
real structure. Because an SVD arranges the dimensions in decreasing orde
of the magnitude of the singular values, the later dimensions with little or n
structure will appear at the end. Hence we have to make a decision only abou
which value of k to use. Parts corresponding to the $k + 1$st and subsequen
singular values can be discarded.

An appropriate choice for k is made by considering the magnitude c
the singular values, which provide a measure of how much variation is bein
captured in each dimension. There are two standard ways to make this choic

and at least two other, more sophisticated, methods that are beginning to
used.

The first standard approach is to plot the singular values using a sc
plot, a plot of the magnitudes of the singular values in order. Since these
non-increasing, and often decrease quite quickly, they resemble the side c
mountain, which is the origin of the name. A suitable cutoff is a value c
where this slope seems to flatten or when there is a detectable elbow or k
in the curve.

The second considers the contribution of each singular value to the wh
in a slightly more formal way. The contribution of each singular value can
computed by

$$f_k = s_k^2 / \sum_{i=1}^{r} s_i^2$$

and then the entropy of the dataset calculated as

$$\text{entropy} = \frac{-1}{\log r} \sum_{k=1}^{r} f_k \log(f_k)$$

Entropy measures the amount of disorder in a set of objects; in this case
has a value between 0 (all variation is captured in the first dimension) a
1 (all dimensions are equally important) [8]. The magnitude of the entre
indicates how many dimensions need to be retained.

*The values of the f_k for our example matrix are: 0.357, 0.227, 0.170, 0.1
0.043, 0.027, 0.01454, and 0.0008871. The entropy for this dataset is 0.7
suggesting that capturing variation requires most dimensions.*

The third method is to use the technique of Zhu and Ghodsi [118] wh
is based on the assumption that the singular values are drawn from two
ferent distributions, one for the significant components and the other for
noise components. An expression for the profile log-likelihood of the choice
k is constructed from the combination of these distributions, and the ma
mum log-likelihood is determined empirically. This maximum corresponds
the best choice of k.

A fourth method is to choose k such that the residual matrix of
$k + 1$ to m components appears to be a random matrix. Suppose a matri
multiplied pointwise by a random $-1, +1$ matrix. Its Frobenius norm does
change. If it is a random matrix, that is it contains no structure, only no
its 2-norm will not change either. However, if it contains structure, alter
the signs of its entries will change the 2-norm by an amount that refle
the amount of structure present. Hence the difference of the 2-norms of
residual matrix and the matrix obtained from it by pointwise multiplicat

by a random $-1, +1$ matrix, divided by the Frobenius norm should becom
small as soon as the residual matrix contains only noise [4].

The truncated SVD is the best representation of the data in the sens
of capturing the variation among the objects and attributes. The matrix A
that results from remultiplying the truncated matrices on the right hand sid
of the decomposition is the best approximation to A in both the 2-norm an
Frobenius norm. Chu [25] has also shown that a truncated SVD is the bes
minimum variance estimation of the random variable corresponding to th
rows; in fact truncation corresponds to minimum variance estimation. Henc
an SVD provides the best representation of the data in a statistical sense a
well.

Removing redundancy

There are other reasons why we might want to discard components beyon
a certain point, even if those we discard are not simply noise. The orderin
of components ensures that the most important structures in the datase
appear in the first few components, less important structure in subsequer
components, and then possibly noise in later components. Truncating a
any value of k preserves as much structure as possible for that number o
dimensions. So, in a sense, the choice of k is not critical since there is a smoot
relationship between the value of k and the amount of structure preserved.

Choosing a small value of k may allow the important structures to b
discovered more easily without the distraction of less-important structure
Furthermore, the distances between points are cheaper to compute in a lowe
dimensional space (requiring only a sum of k terms).

Normally the first k dimensions of the U and V matrices are used fo
subsequent analysis. However, it may sometimes be useful to choose othe
dimensions, and examine the similarity of points in the spaces so defined. Fo
example, the first few components may capture structure in the data that i
already well understood, and it may be the deeper structure that needs to b
analyzed.

Visualization

Of course, there are special advantages when k is 2 or 3 since we can visualiz
the position of the points directly. This allows human abilities to see structur
in visual data to be exploited.

When a larger k is required, helpful visualization can still be achieve
by plotting three dimensions at a time. Some visualization packages contain
display routine called a *Grand Tour* which displays k-dimensional data, thre

dimensions at a time, in a way that helps a human observer to see which
any, dimensions contain interesting structure.

Figures 3.6 and 3.7 show plots of the entries of the matrices U and V tr
cated to 3 dimensions. Figure 3.8 shows the scree plot of the singular val
of this decomposition. In Matlab, these visualizations can be rotated on
screen, making it much easier to see their three-dimensional structure (
Appendix A).

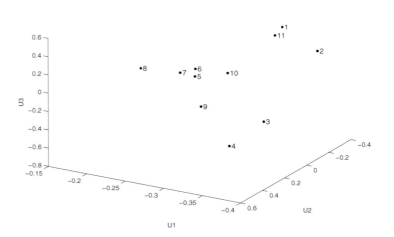

Figure 3.6. *3-dimensional plot of rows of the U matrix.*

Some care needs to be taken when computing distances in the sp
spanned by the k rows of V because distances in the first dimension are m
significant than distances in the second dimension (by exactly the ratio σ_1/c
and so on. Hence it may often be better to use the rows of the product U
as the coordinates – the plot looks the same but the axis lengths are differe
Ordinary distance computations can then be used in this space.

We will call the space spanned by the rows of V, in which the rows o
define coordinates for the objects, U *space*; and the symmetric space in wh
the rows of V define coordinates for the attributes, V *space*.

Figures 3.9 and 3.10 show plots in 3 dimensions of the entries of US and V
The relative positions of the points are unchanged from Figures 3.6 and 3
but the axes have different scales.

Figures 3.11, 3.12 and 3.13 show the same plots for the example ma
normalized using z scores, again truncated to 3 dimensions.

Selecting objects and/or attributes with special properties

The correlation matrices AA' (for the objects) and $A'A$ (for the attributes) provide information about the relationships in the dataset. However, the equivalent *truncated* correlation matrices provide even better information, and in a way that can be related to the SVD [75, 76].

Let A_k be a matrix obtained by multiplying together some k rows of U, the matching elements of S, and the matching rows of V. Consider the correlation matrix $A_k A_k'$, which we might expect to tell us something about the correlation among objects due to the k subprocess(es) that remain. Expanding A_k using the SVD we find that

$$
\begin{aligned}
A_k A_k' &= (U_k S_k V_k') (U_k S_k V_k')' \\
&= U_k S_k V_k' V_k S_k U_k' \\
&= U_k S_k^2 U_k'
\end{aligned}
$$

since $V_k' V_k = I$ and $S_k' = S_k$. So the ijth entry of $A_k A_k'$ is the dot product of the ith and jth rows of U_k, weighted by the squares of the singular values. (In exactly the same way, the entries of $A_k' A_k$ are weighted dot products of the rows of V_k.)

The magnitudes of the entries in the correlation matrix obtained by truncating after the first few singular values provide a good estimate of the

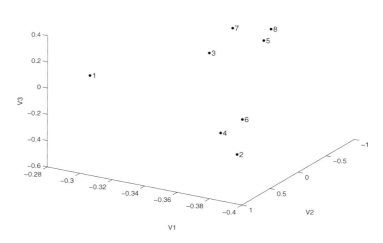

Figure 3.7. *3-dimensional plot of the rows of the V matrix (columns of V').*

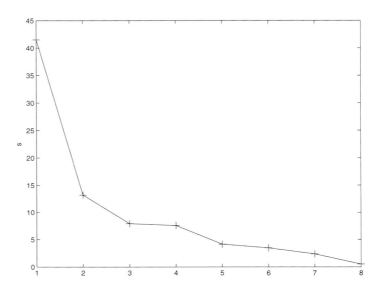

Figure 3.8. *Scree plot of the singular values.*

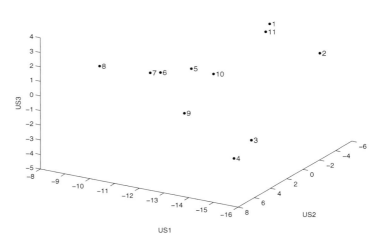

Figure 3.9. *3-dimensional plot of US.*

correlation between objects and/or attributes for the process represented
the choice of k.

Unlike the direct correlation matrix of A, the correlation matrix a

truncation reflects both the absence of correlations in the processes that ha
been ignored (the lost information due to truncation), and higher-order tru
cation information. For example, two objects with no direct correlation ma
have indirect correlations via some other object, and their mutual correlation
with some of the attributes. Matrices such as $A_k A'_k$ may be useful inputs
other analysis techniques since they encapsulate information neatly.

The correlation matrix from the truncated SVD of the normalized version
A, truncated at $k = 2$, is:

$$
\begin{bmatrix}
4.84 & 2.37 & -0.39 & -7.08 & -4.47 & 3.12 & 0.52 & 1.71 & -1.62 & 0.62 & 0.3 \\
2.37 & 1.43 & 1.43 & -3.26 & -3.73 & 2.00 & 0.12 & 0.57 & -0.71 & 0.40 & -0.6 \\
-0.39 & 1.43 & 9.88 & 1.89 & -8.99 & 2.61 & -0.87 & -1.75 & 0.67 & 0.53 & -5.0 \\
-7.08 & -3.26 & 1.89 & 10.53 & 5.28 & -4.18 & -0.87 & -2.72 & 2.45 & -0.83 & -1.2 \\
-4.47 & -3.73 & -8.99 & 5.28 & 13.01 & -5.60 & 0.31 & -0.04 & 0.98 & -1.13 & 4.3 \\
3.12 & 2.00 & 2.61 & -4.18 & -5.60 & 2.85 & 0.09 & 0.63 & -0.89 & 0.57 & -1.2 \\
0.52 & 0.12 & -0.87 & -0.87 & 0.31 & 0.09 & 0.13 & 0.32 & -0.22 & 0.02 & 0.4 \\
1.71 & 0.57 & -1.75 & -2.72 & -0.04 & 0.63 & 0.32 & 0.87 & -0.66 & 0.13 & 0.9 \\
-1.62 & -0.71 & 0.67 & 2.45 & 0.98 & -0.89 & -0.22 & -0.66 & 0.57 & -0.18 & -0.4 \\
0.62 & 0.40 & 0.53 & -0.83 & -1.13 & 0.57 & 0.02 & 0.13 & -0.18 & 0.11 & -0.2 \\
0.36 & -0.64 & -5.01 & -1.20 & 4.39 & -1.21 & 0.46 & 0.95 & -0.40 & -0.25 & 2.5 \\
\end{bmatrix}
$$

The negative correlations between, for example, objects 1 and 2, and objec
4, 5, and 9 become clear from this matrix.

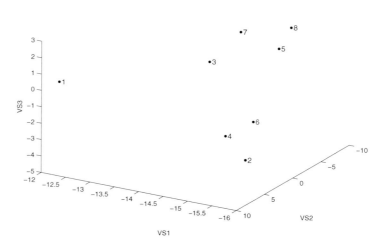

Figure 3.10. *3-dimensional plot of VS.*

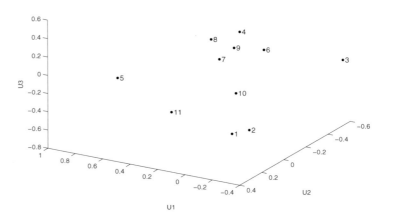

Figure 3.11. *3-dimensional plot of rows of U when the exam dataset, A, is normalized using z scores.*

3.3.2   Similarity and clustering

The main advantage of an SVD is that, under the geometric interpretati truncating the U and V matrices avoids the difficulties of working with met.

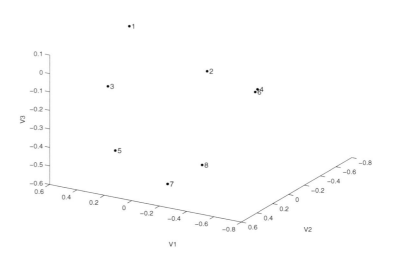

Figure 3.12. *3-dimensional plot of rows of V when the exam dataset, A, is normalized using z scores.*

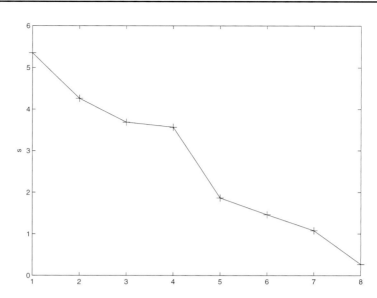

Figure 3.13. *Scree plot of singular values when the example datase*
A, is normalized using z scores.

in high-dimensional spaces, while preserving as accurate a representation i
low dimension as possible.

The two commonest measures of similarity among objects or attribute
are:

- Euclidean distance. Computing the Euclidean distance between a pa
  of points is cheaper ($\mathcal{O}(k)$ rather than $\mathcal{O}(m)$) and more effective tha
  computing the distance between them in the original space.

- Cosine similarity. This measures the closeness of the two vectors fron
  the origin to each of the points and is useful when the appropriate row
  or columns of A have been normalized so that the points are effectivel
  on the surface of a unit sphere; or when the entries in the matrix ar
  sparse and sparsity would be destroyed by normalization.

  This happens, for example, in word-document matrices which are spars
  because most words occur in only a few documents, and where the fac
  that a word occurs at all in a document is more interesting than it
  frequency.

A vast number of clustering techniques based on SVD have been deve
oped. Often this has happened in particular problem domains, and many c
them are not well known outside of these domains. Several have been rein
vented repeatedly. They all rely on taking some subset of the singular vectors

the columns of U for clustering objects and the columns of V for cluster
attributes.

Here are some techniques for clustering:

- Use an ordinary clustering technique on the singular vectors, for exam
  k-means. Applying such a technique to the rows of U (especially a
  truncation) rather than the rows of A exploits the fact that 'noise'
  been removed from the dataset and its dimensionality reduced. So
  ordinary clustering technique should produce a better result and a
  run faster.

- Treat the new axes (the right singular vectors) as cluster centr
  and allocate each object to the appropriate cluster, giving priority
  the right singular vectors in order (and the converse for clustering
  tributes). In other words, the first cluster contains all of those po
  that fall within a $45°$ cone around the first new axis, the second clus
  contains all points that fall within a cone around the second new a
  and so on. The points that fall in the cone around the kth new a
  can be treated as a cluster in the same way as the others, or could
  considered as the 'everything else' cluster, in other words as a set
  outliers.

  Each cone is really two cones, one consisting of vectors positively c
  related with the axis, and the other consisting of vectors negativ
  correlated with it. In some applications, it might be sensible to consi
  objects in both cones as forming a single cluster; in others they mi
  be considered as forming two different clusters.

- Look for 'plateaus' in the first left singular vectors: either by sort
  the values from a column of U (usually the first column) and plott
  them directly [35], or by histogramming their values. This approach
  considerable theoretical support, but it is hard to use in practice beca
  (a) clear plateaus and steps between them do not tend to appear in
  datasets, and (b) in any case the boundaries between such structu
  tend to require subjective choice.

  Alpert [7] shows that such clusterings improve when more columns
  U are used. In fact, it is best to use all r of them. Other papers h
  used various functions of the columns of U as values to be clustered

There are also ways to consider the entries in a matrix as defining
edges of a graph, and then partitioning this graph to cluster the data. T
approach is so important that the next chapter is devoted to it.

3.3.3 Finding local relationships

Normalization of datasets for an SVD means that their data values are c
similar magnitude. Multiplying row(s) or column(s) of the dataset by a scala
effectively changes their influence on the entire decomposition. If the scala
is greater than one, the effect is to move the points corresponding to thes
rows or columns further from the origin. However, this also has the usefu
side-effect of 'pulling' points that are correlated with the upweighted point
further from the origin as well. Furthermore, increasing the weight on som
objects also moves the points corresponding to attributes that are correlate
with these objects.

This property can be used to see clusters of correlated objects and at
tributes that would otherwise be hidden inside larger clusters, provided a
least one such object is known. If this known object is upweighted, then i
will move, and also change the position of correlated objects. All of thes
objects can then be upweighted and the SVD repeated. When new point
stop being moved outwards, the current set of upweighted objects probabl
represents a well-defined cluster.

The same process can be used to determine roughly which attribute
account for membership of a cluster of objects (and *vice versa*). For if in
creasing the weight on the objects in the cluster has the effect of moving som
set of attributes, then increasing the weight on those attributes should hav
the effect of moving the cluster of objects – and this can be checked by th
appropriate SVDs.

Adding artificial objects to orient dimensions

One of the weaknesses of SVD is that the pseudoattributes or dimensions c
the transformed spaces cannot be easily understood because they are linea
combinations of all of the original attributes. However, the significance of th
first few dimensions can sometimes be discovered by adding extra artificia
objects to the dataset representing extremal examples of some property c
interest. For example, if we suspect that the first transformed dimension i
capturing the total magnitude of the attributes associated with each objec
then we can add artificial objects whose total magnitudes are larger than
and smaller than, those of any normal object in the dataset. If the point
corresponding to the artificial objects are at the extremes of one dimensio
in the transformed space, then we can be confident that this dimension i
capturing total magnitude. For example, recall the first dimension of th
transformed wine dataset, with one person who had given all of the wines lo
scores.

*Suppose that we add two extra rows to A, one consisting entirely of 1s an
the other consisting entirely of 9s. The resulting plot of the 3-dimensiona*

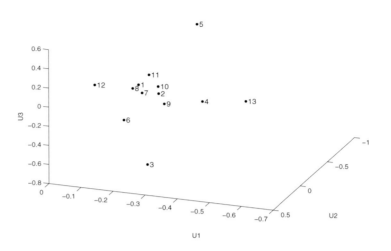

Figure 3.14. *3-dimensional plot of U with a high-magnitude (*
and a low-magnitude (12) object added.

truncation of U is shown in Figure 3.14. The points corresponding to th
new points (numbers 12 and 13) are at opposite ends of U1 dimension, s
gesting strongly that the variation captured by this dimension is that of t
magnitude.

In the same way, we can add rows to the original matrix A, one u
small values in the first four columns and large values in the last four colum
and the other with large values in the first four columns and small ones in
last four columns. The results in this case are shown in Figure 3.15. I
clear from the plot that the second dimension captures the differences betw
objects that have large values in the early columns and those that have la
values in the later columns.

It can also be useful to add lines to the plot of objects, each one indic
ing the direction of the one of the original axes. This can aid in interpretat
in the same way as adding objects to orient dimensions. It can also sl
visually when several attributes are highly correlated.

Figure 3.16 shows the axes of the original space for our example matrix. I
clear that attributes 2, 4, and 6 are very similar, as are attributes 1, 3, 5,
7.

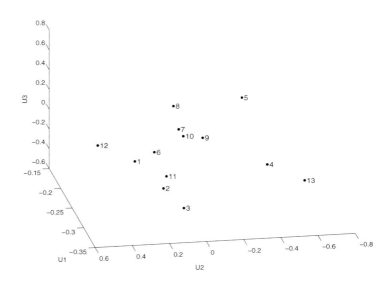

Figure 3.15. *3-dimensional plot of U with two orienting object added, one (12) with large magnitudes for the first few attributes and sma magnitudes for the others, and another (13) with opposite magnitudes.*

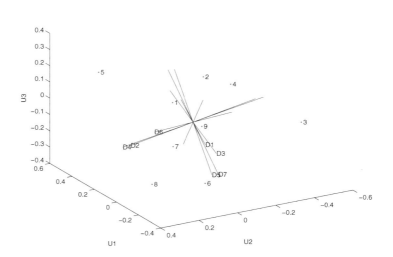

Figure 3.16. *3-dimensional plot of U with lines representing axe from the original space.*

Using the Split V technique

Matrix decompositions are usually applied to datasets that do not hav
target attribute. However, if a target attribute is known, or if we are interes
in investigating how one particular attribute is affected by the others, S
can provide some insight.

If the target attribute takes two values (a situation we can create
an arbitrary attribute by choosing a midpoint value), then the matrix can
divided into two parts: one associated with one value of the target attribu
and the other with the other value. The SVD of each of these matrices p
duces two V matrices, say V_1 and V_2. If the points from these matrices
plotted in the same space, then the different positions of each attribute in
and V_2 give an indication of how the attribute interacts with the two val
of the target attribute. Attributes that move a long distance from one plo
the other tend to be good predictors of the target attribute.

This technique implicitly assumes that the two submatrices are reas
able samples from some larger universe of data and so their SVDs can
plotted in the same space. This assumption may not be valid for particu
datasets.

3.3.4   Sampling and sparsifying by removing values

Many fast approximate algorithms for computing low-rank approximation:
the SVD are known. These are of interest from the practical point of v
of the resources required for what is otherwise an expensive algorithm;
the existence of such algorithms also reveals something about the proper
required in datasets.

The matrix A_k can be computed in time independent of n and m w
high probability [45] under some plausibly practical assumptions. Howe
constructing the matrix requires time $\mathcal{O}(kmn)$.

Entries of the matrix A can be randomly omitted or quantized (say
0 and 1) without affecting the linear structure of the matrix and so with
affecting the SVD [4]. These changes to the dataset matrix amount to addi
matrix of independent random variables with zero mean and bounded varia
to it. Such variables have no linear structure and so 'vanish' from an SV
This result shows that SVD has some ability to ignore random backgrou
which is useful when the goal is to find small pockets of correlation in larg
uncorrelated datasets. These fast algorithms work well when the goal of
analysis is to understand the mainstream structure in the data, but should
used with caution when the goal is to understand finer structure.

3.3.5 Using domain knowledge or priors

The following result shows how to take a matrix of rank m and reduce its rank one step at a time. The reduction process produces a matrix decomposition that is quite general. Suppose that A is an $n \times m$ matrix, x is an $n \times 1$ vector and y is an $m \times 1$ vector such that

$$\omega = x'Ay \neq 0 \tag{3.1}$$

Then the matrix

$$B = A - \omega^{-1}Ayx'A$$

has rank exactly one less than the rank of A [26], a result due originally to Wedderburn.

Applying the theorem once reduces the rank of the dataset matrix, A by one, and produces a vector, x, that can become the first column of the left hand matrix of a decomposition, which we have been calling C. Similarly, the vector y becomes the first row of the right-hand matrix of a decomposition which we have been calling F. The theorem can be applied repeatedly. After each round, the rank of the dataset matrix has been reduced by one, a new column has been added to C, and a new row has been added to F.

The theorem allows a matrix of rank m to be decomposed in many ways, depending on how x and y are chosen. If the xs and ys are chosen to be orthogonal, then the result is an SVD. However, the choice of the first x and y can be made freely, subject to condition (3.1). A standard decomposition algorithm can then be applied, starting from matrix B. These initial choices of x and y can be used to include external information in the decomposition. For example, x could specify some subclass of objects by putting 1s in the corresponding positions of the vector, and 0s in the other positions. This provides a mechanism to include domain knowledge or priors in a matrix decomposition.

3.4 Algorithm issues

3.4.1 Algorithms and complexity

The complexity of SVD is $n^2m + nm^2$. Since m is typically much smaller than n in data mining algorithms, the complexity is usually taken to be $\mathcal{O}(n^2m)$. In data-mining applications, n is often large, so computing the SVD is expensive.

Many computational packages (for example, Matlab, Octave, R, S) contain a command to compute an SVD. Standalone software in most programming languages is also readily available.

3.4.2   Updating an SVD

An SVD can be updated in two senses. The first is that the matrix A
mains the same size, but has had some of its values changed. In this ca
it is straightforward to recompute the SVD incrementally [88, 105]. The ti
complexity is linear in n provided that the magnitude of the changes are sm

An SVD can also be updated in the sense that new rows or columns
added. Rearranging the SVD equation we see that

$$U = AVS^{-1}$$

Hence given a new row of A, whose shape is $1 \times m$, this equation can
applied to transform it to a new row of U, whose shape is $1 \times r$. A sim
procedure can be used to update V. This computation is not a true upd
since it does not preserve orthogonalities, but it is cheap. If desired,
previous incremental algorithm can be run on the new matrices to reinfo
the orthogonality.

3.5   Applications of SVD

3.5.1   The workhorse of noise removal

The number one application of SVD is noise removal. Typically experimen
data has been collected, and an SVD is applied to determine how noisy it
and perhaps to remove the noise component. Often, this is all that is do
none of the other benefits of SVDs are used.

3.5.2   Information retrieval – Latent Semantic Indexing (LSI)

SVD has been used extensively in information retrieval, where it is known
latent semantic indexing [13, 32].

In any large information repository, one of the critical problems is fi
ing documents about a particular topic. One approach to this informat
retrieval problem is to model each document as a vector, with one entry
each possible word. If a particular word is not in the document, then
corresponding element is set to 0; if the word is present, then the correspon
ing element can either be set to the number of times the word occurs
the document, some modification of this frequency that reflects the fact t
longer documents contain more words, or just to 1 to indicate that the wo
is present. The entire repository is then modelled by a matrix with one n
for each document, and one column for each word.

The problem is that there are many possible words, even if stemming is used (so that, for example, 'skip', 'skipping', 'skipped' are all treated as the same word). If the information repository is very large, then the matrix will have many rows; but no matter how large the repository, the matrix will have many columns. For example, a typical collection of documents such as the back issues of a newspaper can easily mention more than 100,000 words.

The fundamental problem in information retrieval is: given some search terms, retrieve all of the documents that contain those search terms or, perhaps more usefully, return documents whose content is semantically related to the search terms. For example, if one of the search terms was 'automobile' it might be appropriate to return also documents that contain the search term 'car'.

It is also usually important to retrieve the documents and arrange them in some order of importance, so that the most relevant documents come first. This is less important when there are typically only a few relevant documents that match a set of search terms, but becomes the most important part of the process when the number of relevant documents is large, as it usually is in web search.

A search query can be thought of as a short pseudodocument that contains only the search terms, and so can be described by a row of the same form as the rows of the document-term matrix. The goal of search is then to find rows in the document-term matrix that are similar to the search term row, where similarity means 'contains all of the words'.

Progress was made with this problem by using the geometric model, and treating documents as vectors in a very high dimensional space, a search vector as a new vector in the same space, and retrieving document vectors that are close to the search vector. The measure of similarity used is cosine similarity – that is vectors are close if they point in the same direction in this space – because a query always looks like a very *small* document, no matter how normalization for document length is computed, so direct Euclidean distance is not a good metric.

There are several problems with vector-based retrieval. First, the space is very high dimensional, which creates problems for distance measurement. Second, it treats each word as independent, whereas in languages like English the same word can mean two different things ('bear' a burden versus 'bear' in the woods), and two different words can mean the same thing ('car' and 'automobile').

Rather than implementing information retrieval in the space described by A, it is effective to compute the SVD of A, and implement information retrieval in the space described by U, truncating it at some appropriate value of k, perhaps $k \sim 10,000$.

This has a number of obvious benefits. First, there is a great dea$l$ redundancy in natural language, and this redundancy is largely eliminated truncation.

Second, words that are synonyms tend to be placed close together in truncated space, a nice illustration of SVD's ability to exploit higher-order formation. Consider two similar words such as 'car' and 'automobile'. The words tend not to co-occur in the same document, since they mean roug$h$ the same thing and the choice of which to use is stylistic. The columns the raw data matrix corresponding to these two words are therefore co pletely different; and the rows corresponding to documents about 'car's at least slightly different to those about 'automobile's. However, the context of occurrences of 'car' and 'automobile' will be similar, so these two set$s$ rows will have similarities because of other, contextual words. SVD is able to transform this similarity of rows into similarities between the 'car' and 'automobile' columns in the truncated representation. This has an important payoff: a query that is expressed using the term 'car' will also find documents that contain occurrences of 'automobile' but *none at all* of 'car'. The region selected by the presence of the term 'car' is (close to) the region selected 'automobile', so these search terms have become approximately interchange able. To say it another way, the truncated space has generalized beyond both specific terms to a deeper concept.

Third, words with more than one meaning are placed between region corresponding to their different meanings – they are pulled toward such region with forces that reflect how common each meaning is. This has the effec$t$ discounting the usefulness of such words as search terms, since they do select any particular region strongly.

This approach is so effective that it can even be used for cross-lingual information retrieval, where the search terms are in a different language from the retrieved documents. A document space containing some documents both of the languages is required, but documents in only one language also permitted. Queries in one language retrieve relevant documents in other language (where 'relevant' means that they have the same underlying semantics) [38].

Document-word matrices are typically extremely sparse – any given doc ument contains only a relatively small number of different words out of possible 100,000+. LSI is able to discover and measure the similarity between documents that do not share even a single word, and similarities between words that never co-occur in a single document.

3.5.3   Ranking objects and attributes by interestingness

In a truncated SVD, there are only k different directions in which vectors ca
point and be orthogonal, whereas there are n vectors corresponding to th
rows of U. There are two kinds of vectors that end close to the origin. Th
first are objects that are correlated with many of the other objects. Suc
vectors would like to point in every direction so that they can have positi
dot products with many other vectors. Or, using the spring intuition, the
are pulled from every direction and so end up near the center. The secon
kind of vectors that end close to the origin are those that correlate with nor
of the objects. These vectors want to be orthogonal to all of the other vector
and so must also have small magnitudes.

This means that vectors whose ends are far from the origin correspon
to objects whose correlation with all of the other objects is unusual. Suc
objects are interesting in a meaningful sense. For example, we could ran
objects based on their distance from the origin; those objects at the top c
the ranking are those with interesting correlation with the other objects, whi
those at the bottom of the ranking are uninteresting in the sense of bein
correlated with almost all of the other objects, or with very few of them. (C
course, using only distance from the origin loses information about directio
which is also important.)

This technique can be used in many settings. For example, given
document-word matrix it can be used to rank documents in a way that s
lects the most interesting documents. Very bland documents, those that us
common words in common ways, are pushed to the bottom of the ranked lis
So are documents that use only rare words. Those documents that appea
near the top of the ranked list are those that use moderately common word
but in ways that are different from the other highly-ranked documents. Th
same approach can be used to rank words in a document repository, rankin
both common words and uncommon words low, while selecting words tha
have interesting correlations. Such words are likely to be useful terms fc
indexing or forming a taxonomy for the repository, since they all have usef
discriminatory power for the documents. This approach of ranking was use
in [101] to detect groups of messages in which ordinary words had been r
placed by other words of different natural frequency in an attempt to conce
the content of emails.

3.5.4   Collaborative filtering

Recommender systems use information about a set of objects and their prope
ties to predict new properties of a particular object. Most of their application
have been in the commercial sector, such as recommending movies or book
For example, a bookseller such as Amazon has a large number of record

one for each purchaser, listing the books purchased. If a customer come
their web site, they can recommend a book that the customer might like,
considering the patterns of other, similar customers' book purchases.

There are several ways to implement such recommendation syste
One way is to describe each book by a short list of properties. Each custo
can then be described by merging the lists of properties of books he or
has purchased. A good book to recommend would be one that has a sim
list of properties to that of the customer. This approach has two drawba
each book's properties have to be described, perhaps by a human, altho
for some objects this information can be extracted automatically; and no
is made of the purchasing information of other customers. Such systems
called *content recommenders*.

Collaborative filtering tries to avoid these drawbacks by using infor
tion about purchase patterns of other customers. There is no need to explic
understand what properties of a book make it attractive to a particular s
set of customers; it is enough to be able to identify such a subset. A
information about which customers liked which books provides more su
information about preferences than a short list of properties can do.

The simplest form of collaborative filtering requires a matrix whose r
correspond to customers and whose columns correspond, say, to books.
entries in the matrix can either be zeros and ones, indicating whether
not customer i purchased book j, or could be some indication of how m
customer i liked book j, perhaps expressed on a scale from 1 to 10. Gett
purchase information is free. Getting information about how much a book
liked is more difficult because the customer has to provide extra informat
and such opinions tend to vary widely over short time periods, even for
same customer and book.

If we wish to provide a recommendation for customer k, then we sim
need to find a row i that is similar to row k, and look for a column where
i has a non-zero entry and k has a zero entry. The book corresponding to
column is the one that should be recommended. (Note the similarity to
information retrieval problem.)

There are several obvious problems with this simple scenario. Fi
row i might be most similar to row k, but still not contain much over
that is be closest, but not actually very close. The matrix is very sparse
most people will not have purchased most books. Second, there may be m
possible recommendations, and it is not obvious how to select the best c
One approach is to select the book (= column) that has the most entr
but this recommends popular books at the expense of books that might
a better fit. There are also problems that mimic the problems with word
the previous section: there are often different books with the same title an
may be hard to tell them apart; also many authors write books in series,

the books are very similar. However, this simple algorithm treats all book
as distinct entities, and so cannot use this extra structure. For example,
you buy a book by a particular author from Amazon, you immediately g
recommendations for all of that author's other books, even though it is likel
that you already own or have read many of them. Even if you tell Amazo
that you own many books by the same author, it persists in recommendin
the remaining ones.

Collaborative filtering can be applied to many other settings, charac
terized by using small amounts of human input about quality or preference
to compute either global or local rankings of objects. The book system
easily extended to providing recommendations for movies, or restaurants, c
music. The same approach can be used to provide feedback on technical pa
pers, or news articles (e.g., Slashdot), or publish-subscribe or RSS system
which match content, as it is created, with a set of user preferences that ma
themselves be either static or dynamic.

The performance of collaborative filtering can be improved by using SV
to decompose the person-object matrix. People don't prefer a random selec
tion of different books; they have preferences for certain types of books, c
books by certain authors. (Physical) libraries use this information to organiz
books into different categories, for example mystery, or science fiction, or ho
ror; and organize non-fiction into one of a set of well-defined taxonomies. Th
is helpful for someone whose tastes fall cleanly within a category. Librarie
also shelve fiction books by author. Again this is helpful to someone who like
a particular author as long as the author does not write under several name

Collaborative filtering systems can organize books into clusters that re
flect how a large number of people actually prefer them, rather than a tax
onomy imposed from outside. Clustering the rows of the U matrix avoi
many of the drawbacks of working directly with the person-object matri
Redundancy in the data has been removed; the main dimensions along whic
books are categorized can be seen; books that appear to be different but ar
purchased by the same people are placed close together, and different book
with the same name are placed far from the clusters in which they would, ir
dividually, fall. The result is a space in which comparisons between customer
can be made more reliably.

The U matrix, when truncated at some suitable k, is a much smalle
matrix than the original person-object matrix, and it is also much less spars
When we want to find customers similar to a particular customer, we ar
working with distances that are better behaved because they are calculate
in a lower-dimensional space.

When we find a customer who is similar to the customer of interes
we cannot directly produce a recommendation because the columns of the l
matrix do not correspond to books, but rather to groupings of books. Ther

are a number of ways to compute the desired recommendation. One wo
be to change the values of the customer of interest to move that row halfv
towards the row of the customer whose recommendations we plan to use. T
new row can then by mapped from U-space back to the space of the origi
person-object matrix and compared to the original row. The entry that
increased the most could suggest the best book to recommend. Alternativ
a selection of books could be recommended, ordered by how much their entr
have increased.

There are interesting possibilities for folding in other business conside
tions, for example the amount of profit associated with each book. Weight
each column by a scalar corresponding to the desirability of selling each bo
biases the recommendations in favor of such books. The preference inforr
tion from other customers is still used, but is modulated by another kind
information.

We can use the ranking described in the previous section to analyze r
ommendation data further. The ability to rank the rows of the matrix by h
interesting they are selects those customers whose rankings are the most u
ful. Those who buy only what everyone else buys, for example bestsellers,
ranked low. So are those whose purchases are idiosyncratic. Those custom
who appear at the top of the ranking from the SVD are those whose ranki
are most helpful in a global sense. They can be encouraged by providing eit.
a psychic benefit (publishing their names), or an economic benefit (disco
coupons) to encourage them to continue to provide high-quality informat.
for the system. Notice that there is an inherent bias in these kinds of s
tems: there is usually a reasonable motivation to provide positive informat.
– those who liked a book would like others to enjoy it too. However, ther
less motivation to provide negative information, and what motivation ther
perhaps is less trustworthy.

The ability to rank the columns, for example books, provides anot
new kind of useful information. Booksellers understand the economics
bestsellers. They are also coming to understand the economics of the 'lc
tail', keeping low-demand books available for rare customers. However,
books or other objects that appear at the top of the ranking from an S'
are those books that attract interesting customers. These books repres
potential word-of-mouth bestsellers [47], and so identifying them represe
a huge economic opportunity. Consider the 'usual' ranking based on sales
expected revenue, and imagine that it is bent into a U shape. The part of
list that comes at the top of the ranked list from the SVD is the part fr
the bend of the U, neither the bestsellers, nor the rare sellers, but those
between. Such books are neither correlated with very popular books (b
sellers), nor uncorrelated with almost all other books (rare sellers). T
correlation information about books is derived, indirectly, from custome
Amazon, for example, could use this to make more sophisticated predictio

There may also be some benefit to understanding the clustering of book implied by V matrix. This should agree with more conventional clustering but the differences might be revealing. For example, they might suggest ne ways of thinking about groupings of books; perhaps "detective fantasy" becoming a new category.

The effects of normalization can be clearly seen in the collaborativ filtering setting. Suppose that participants are asked to rate books on a sca from 1 to 5, with 1 meaning 'didn't like it' and 5 meaning 'liked it ver much'. The data will typically be very sparse, since few people will have rea more than 1% of the books in a typical collaborative filtering setting. Th percentage might be higher in other settings, for example movies and musi but it is probably never very large.

So we have a matrix with rows corresponding to people and columr corresponding to books. If we do not normalize the data at all, then the firs component of the attribute space will capture the average approval rating fo all books. Bestsellers will be far from the origin because they will have hig ratings from many people, but books that were strongly liked by moderatel sized groups of people cannot be distinguished from books that were consic ered mediocre by many people. Also the orthogonality requirement of an SV means that the direction of subsequent components will be distorted by th first component.

Normalizing using z scores and including all of the zero values in th normalization introduces a skew into the data. Suppose that all of the ra ings are positive. The zero entries are reflected in the denominator of th calculation of the mean, so that the effect of the non-zero ratings are heav ily damped by the zeros, and the value of the mean will typically be quit small. Subtracting the mean from all of the column entries means that th previously zero entries will all become slightly negative. Hence the resultin distribution of values will be biased slightly towards the negative direction fo every attribute – although the original zero entries represent no informatior their effect is to skew the data that will be analyzed.

A better approach is to leave the zero entries unchanged, and adjust th non-zero entries, whose range of values are known, by centering them aroun zero. This maps 5 to +2 and 1 to -2, making a rating of 1 antithetical to rating of 5. This is appropriate, and may even make it easier to distinguis books that are generally liked from those that are generally disliked. Rating of 3, which suggest a neutral opinion about a book, have now been mapped t 0. The critical side-effect of this normalization is that neutral opinions about book have been conflated with the absence of opinions about a book. In a wa this is reasonable; it assumes that both provide no extra information about th book's quality. However, a book that many people have read and rated as a really is a mediocre book, while a book that has not been read by anyone is a unknown quantity, and both will look the same. The normalization prevent

us from knowing where, on this considerable spectrum, a particular b
actually lies. In the end, this normalization forces the matrix decomposit
to ignore information about mediocre ratings and transform the data ba
on the strong positive and negative opinions.

3.5.5   Winnowing microarray data

Microarrays are high-throughput devices that measure the expression leve
various proteins, usually messenger RNA (mRNA), in a cell. Each microar
is a slide on which a large number of spots have been placed. Each s
consists of many identical strands of oligonucleotide, complementary DN
or some other similar strand. When a sample is washed over the microar
some of its contents binds to the spots, and the quantity that binds can
detected by subsequently reading the slide. There are many variations in t
and process of collecting microarray data.

Each sample produces an intensity reading for each spot, reflecting h
much the corresponding mRNA has been expressed in the cells of the sam
A typical microarray may contain more than 20,000 spots. A dataset t
might be used to study a particular disease produces an array with, s
20,000 rows and a set of columns, each one corresponding to a single sam
(patient).

The goal of microarray analysis is to discover how an external condit
(having a disease) correlates with an internal situation (increased expres
levels of some genes and decreased levels of others, as reflected in the mR
they express). If there are samples from both the normal and disease cor
tion, then we expect to see differences, for some rows, between the colun
associated with the different samples.

One of the problems is that the differences associated with most cor
tions are expected to affect only some fraction of the 20,000 or more possi
genes (spots). Before carrying out some sophisticated analysis, it might
helpful to remove those rows corresponding to genes whose expression le
do not change in significant ways.

A standard way to do such winnowing is to discard those rows wh
expression levels are low, perhaps aiming to reduce the number of rows b
third. However, this kind of simple winnowing does not do the right thing
retains a single gene that shows a large change in expression level, altho
it is unlikely that most conditions can be explained by a single gene; an
removes genes with small changes in expression level, even when these chan
are highly correlated, potentially missing important, but subtle, express
patterns.

SVD provides a way to winnow such data more appropriately [66]. Gi
a data matrix, A, sorting the rows of U by distance from the origin selects

genes with the most interesting expression in exactly the right sense. Gene whose expression levels do not change across the patient groups will ten to be close to the origin, both because their rows will be almost constan and because there are many other rows like them. Similarly, rows containin unusual expression patterns that appear nowhere else will also tend to be clos to the origin.

A further advantage of creating a ranked list is that the decision abou which rows to retain and which to remove can be made after the winnowin rather than before. Any approach that requires the boundary to be define before the data is examined is much harder to use because information t make a good choice is not known when the decision has to be made.

3.6 Extensions

3.6.1 PDDP

Boley's *Principal Direction Divisive Partitioning* (PDDP) [14] uses SVD t build an unsupervised decision tree. The approach is as follows:

- Compute the SVD of the data matrix, and consider the direction of th first singular vector.

- Partition the objects depending on their position along this vector (ther are several possibilities for the splitting hyperplane which is, of course normal to the first singular vector). The resulting two parts of th original datasets are separated.

- Continue the process on each of the partitions separately.

The result is a binary tree that divides the data in ways that reflect the mos important variation first.

Matlab software for PDDP is available from the web site `www.cs.umn.ed` `~boley/Distribution/PDDP.html`.

3.6.2 The CUR decomposition

There are two situations where the properties of an SVD make it difficult t use on real data. The first is where the data is sparse; the decompositio results in matrices that are no longer sparse, causing storage and analys problems. The second is when rows and columns that are linear combinatior of the original data values do not make sense in the problem setting. Ofter this is because some attributes are allowed to take on only certain values, sa integer values, and the rows and columns of the decomposition take on ot values.

The CUR Decomposition is designed for such situations – it compu a decomposition of the dataset matrix, using *actual* rows and columns the dataset, but at the expense of a less faithful representation (although the end only less faithful by a multiplicative error factor). Hence the C decomposition provides a high-quality excerpt or sketch of the dataset, rat than a new representation.

The CUR Decomposition of a dataset matrix, A, is given by

$$A = CUR$$

where A is $n \times m$, C is a set of c columns of A, R is a set of r rows of and U is a $c \times r$ matrix. Let k be a scalar smaller than the rank of A, choose a multiplicative error, ε. Then a randomized algorithm that choos columns of A, and r columns of A, where c and r are large enough functi of k and depend on ε, and the columns are chosen in a clever way, produ C, U, and R such that

$$||A - CUR||_F \leq (1 + \varepsilon)||A - A_k||_F$$

The computation has about the same complexity as SVD and requires passes over the dataset.

The trick behind this, and other similar decompositions, is to select columns (resp. rows) in a special way. The columns must be chosen so t they form a set of independent and identically distributed variables (wh is just a careful way of saying that they must be chosen randomly), they must be chosen with replacement, that is the probability of choosin particular row *this* time is unaffected by whether it has been chosen befo For the present algorithm, the probability of choosing a row derives from SVD of the dataset matrix, truncated at k. This suffices to ensure that, w high probability, the structure that remains in C and R approximates structure of A sufficiently well. The reason that this kind of approach wo at all is that, almost by definition, a matrix that is actually low rank, doesn't look as if it is, contains lots of repetition or almost repetition.

The right-hand side of this decomposition acts as a kind of sketch the original dataset, but is much, much smaller and so easier to work w in many practical ways. Because the matrices C and R consist of colur and rows from A, they inherit properties such as sparseness. Also beca each column and row *is* a member of the dataset, it must be a reasona element, no matter what kind of constraints apply to such elements. He the CUR Decomposition has many of the attractive properties of SVD, avoids some of its deficiencies as well. However, something has also b lost – the CUR Decomposition describes only the mainstream structure

the dataset, and so cannot be used to understand the structure of outlier
interesting, or borderline objects.

The two main applications of this kind of decomposition so far hav
been:

- Lossy compression. The right-hand side of the decomposition can t
  much, much smaller than the dataset matrix, and yet does not lose muc
  of the information. This is especially true for a sparse dataset matri:
  where an SVD coding would require more storage space because th
  matrices involved would become dense.

- Creating large datasets based on small amounts of data. There ar
  several important settings where the contents of C and R are know:
  these values can be used to construct an approximation to a much large
  matrix. For example, suppose an organization wants to understan
  how to apply incentives to encourage their customers to buy certai
  products. They can use a few of their customers to get preference dat
  about all of their products, creating an R matrix. They can look a
  user purchases based on a few products for which they have alread
  provided incentives, creating a C matrix. Combining these two matric
  extrapolates the user-product information to all combinations of use:
  and products, suggesting ways to target incentives.

Decompositions of the CUR kind are new, and there are no doubt other way
waiting to be discovered, to apply them for data mining.

Notes

The singular value decomposition has a long history, both as a matrix pr
processing technique and as a data analysis technique [48, 63, 109]. Beltran
and Jordan essentially discovered it independently in 1873 and 1874, respe
tively. A good historical survey of SVD and other matrix decompositions ca
be found in Hubert *et al.* [57].

SVD is usually presented as the extension of eigenvectors and eigenvalue
to rectangular matrices. Although we will make this connection in the ne
chapter, it does not seem to be necessary, or even helpful, for effective use o
SVD for data mining.

The wine dataset is discussed further in [65]; I am grateful to Mary-Ann
Williams for providing me with the data.

A good example of the use of SVD in recommender systems is Sarwa
et al. [96].

Chapter 4

Graph Analysis

4.1 Graphs versus datasets

In the previous chapter, we considered what might be called attributed data: sets of records, each of which specified values of the attributes of each object. When such data is clustered, the similarity between records is based on combination of the similarity of the attributes. The simplest, of course, is Euclidean distance, where the squares of the differences between attributes are summed to give an overall similarity (and then a square root is taken).

In this chapter, we turn to data in which some pairwise similarities between the objects are given to us directly: the dataset is an $n \times n$ matrix (where n is the number of objects) whose entries describe the *affinities* between each pair of objects. Many of the affinities will be zero, indicating that there is no direct affinity between the two objects concerned. The other affinities will be positive numbers, with a larger value indicating a stronger affinity. When two objects do not have a direct affinity, we may still be interested in the indirect affinity between them. This, in turn, depends on some way of combining the pairwise affinities.

The natural representation for such data is a graph, in which each vertex or node corresponds to an object, and each pairwise affinity corresponds to a (weighted) edge between the two objects.

There are three different natural ways in which such data can arise:

1. The data directly describes pairwise relationships among the objects. For example, the objects might be individuals, with links between them representing the relationship between them, for example how many

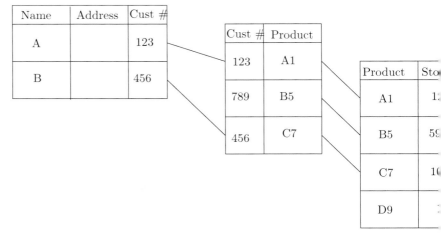

Figure 4.1. *The graph resulting from relational data.*

times they have met in the past year. This kind of data is comn
in *Social Network Analysis.*

2. The data comes from a relational database. The objects are rows fr
   tables, and rows have an affinity if they share a value for a field (t
   is, an affinity represents two entries that would be matched by a jo
   See Figure 4.1 for a small example.

3. The data is already in a geometric space, but it would not be appro
   ate to analyze it, for example to cluster it, directly. The dataset r
   appear high-dimensional but it is known from the problem domain t
   the data actually occupies only a low-dimensional manifold within
   For example, image data can often be very high-dimensional, with
   attribute for each pixel; but the objects visible in the image are c
   three-dimensional, so the degrees of freedom of objects in the scene
   much fewer than they appear.

   Most clustering algorithms have some kind of bias towards convex c
   ters, and so do not perform well when the low-dimensional space
   embedded in the high-dimensional space in an overlapped or contor
   way. It may be more effective to extract an affinity graph based
   local or short-range distances, and then map this graph back int
   low-dimensional geometric space.

   We would like to be able to analyze such datasets in the same way
we did in the previous chapter, but also in some new ways made possible
the fact that the data describes a graph. Some analysis possibilities are:

- *Clustering*: Just as we did for datasets in the previous chapter, we woul
  like to be able to cluster the nodes of the graph so that those in eac
  cluster are similar to each other. This corresponds to finding regions o
  the graph whose nodes are better connected to each other than they ar
  to the rest of the graph or, equivalently, to cutting some graph edge
  that somehow weakly connect the nodes at their ends. Usually, it
  also important that the clusters are of significant size – it is not ver
  interesting to separate almost-isolated nodes from the edge of a grap
  and call them a cluster.

- *Ranking*: We also saw how useful it is to be able to rank objects i
  the previous chapter. For graph data, ranking has to somehow respec
  the affinity structure, so that two nodes that are well-connected to eac
  other should receive similar ranks.

- *Calculating global properties*: Sometimes the global structure of th
  graph is revealing, and this can be described by a few parameters. Fo
  example, given the connections inside an organization, it may be poss
  ble to tell whether decision making is autocratic or democratic, base
  on the 'shape' of the connection graph. It may be possible to determin
  who holds the power in an organization by how central they are in th
  graph. Global properties like these have been much studied in socia
  network analysis. It may also be of interest to know how many co
  nected components the graph breaks into; this tells us whether there
  a single community, or multiple communities present in the data.

- *Edge prediction*: Given the existing edges of the affinity structure, whic
  pair of unconnected edges could be connected by a (weighted) edge mos
  consistently with the existing affinities? This is one way of looking a
  collaborative filtering – from a graph perspective, a recommendation
  implicitly a new edge.

- *Nearest interesting neighbor*: This is really a variant of edge predic
  tion, expressed locally. It's obvious which is the nearest neighbor o
  a given node – the node that is connected to it by the edge with th
  largest weight. However, in some datasets, again especially those use
  for collaborative filtering, nodes with large affinities are near neighbor
  of almost *all* of the other nodes. It may be more useful to find node
  that are similar once this global structure is discounted.

- *Substructure discovery*: Sometimes it is the existence of particular sub
  graphs within the graph that is of interest. For example, money laun
  dering typically requires particular patterns of connection between, sa
  drug dealers, certain kinds of businesses, bank accounts, and people wh
  move money around. It may be useful to be able to discover all occur
  rences of such patterns inside a graph, or all patterns that are unusua
  or some combination of the two.

Even clustering turns out to be more difficult for affinity or graph d than it was for attributed data. Partly, this is for a reason alluded to alrea on page 42; in a geometric space, the distance between any two points depe: only on where each of the points is in space. This provides a number shortcuts when we try to understand the global structure of the data in s a space.

In a graph space, the distance and relationship between two obje depends on all of the other objects that are 'between' them. The additior removal of a single object can alter all of the other, longer-range distanc and so all the quantities that depend on them. Unsurprisingly, algoritl of exponential complexity are required to compute many of the propertie interest exactly.

A general strategy is used to avoid this problem. Rather than w directly in the graph space, various embeddings are used to map the obje and edges into a geometric space, for example a Euclidean space, in suc way that:

- Each pair of connected objects is mapped to a pair of points in sp whose closeness accurately reflects the affinity between them. W affinities are large, separations are small.

- Pairs of objects that are not directly connected are mapped in suc way that their geometric closeness reflects, in some sensible way, tl edge-based closeness in the graph.

The second requirement requires a substantive choice, since there are sev ways in which longer-distance closeness in the graph could be defined, . choosing different ones will obviously make a great difference to the appar properties of the graph.

There are a number of ways in which local affinities can be exten to non-local affinities. The standard graph-theory view is that the dista between two non-local nodes is simply the length of the shortest path betw them (extended to the path with the minimal sum of weights, for a weigh graph). This extension is natural when the edges represent 'steps' with e step increasing the dissimilarity between the nodes. However, in other setti it is natural to consider two non-neighboring nodes to be similar if they connected by short paths and *also* by many different paths. This extensio natural when similarity can be thought of in terms of the 'flow' or (inve 'resistance' between nodes. However, this extension is harder to work v because it requires more of the context of the pair of points to be conside to evaluate their similarity. In fact, two points might be connected by pe through every other node of the graph, so calculating their similarity amou to making a global computation on the graph.

The extension of local affinities to non-local affinities can be even mor
complex when the edge structure of the graph plays a different role to the ro
of the weight structure. We have already discussed collaborative filterin
Collaborative filtering data can be interpreted graphically in a natural wa
– the preference expressed by an individual for a product is a weighted edg
between the two nodes that represent the individual and product. Howeve
an individual who expresses many preferences does not have *better* opinion
just *more* opinions. Nevertheless, the effect of their presence in the graph
to alter the non-local affinity structure of the rest of the graph by providin
short paths that join almost every other individual to almost every othe
product. Clearly, such individuals distort the medium-scale structures fc
other individuals. In the end, this has the disastrous side-effect of makin
the system recommend the most popular products to everyone. This exampl
shows that, in some situations, an affinity extension needs to be even mor
sophisticated than using length of paths and numbers of paths.

We will describe a number of ways to embed a graph space into a geo
metric space. The main difference between them is precisely the issue of ho
they extend pairwise affinity to distance in the geometric space.

4.2 Adjacency matrix

The pairwise affinities between objects define a graph whose nodes or vertice
are the objects and whose edges are the pairwise affinities. The easiest an
most direct representation of these affinities is an *adjacency matrix*.

Given a set of n vertices (corresponding to objects), the adjacency ma
trix, A, is an $n \times n$ matrix whose entries are zero, except that when object
is connected to object j by an edge, the entry has value 1. The matrix ha
n^2 entries and usually the data will describe relatively few pairwise affinitie
so the adjacency matrix will usually be very sparse. Formally, the adjacenc
matrix is

$$A_{ij} = \begin{cases} 1 & \text{object } i \text{ has some affinity to object } j \\ 0 & \text{otherwise} \end{cases}$$

Since we regard affinities as symmetric (the affinity between object i an
object j is the same as that between object j and object i), A is also
symmetric matrix with non-negative entries.

The *degree* of each vertex or object is the number of edges that ar
attached to it, which is the sum of the number of 1s in its row (or equivalentl
column). So the degree of object i is

$$d_i = \sum_{j=1}^{n} A_{ij}$$

The adjacency matrix, as defined so far, takes into account whet or not two objects are directly joined, but does not take into account magnitude of the affinities. We can easily extend it to a weighted adjacen matrix whose entries are weights derived from the affinities, like this

$$A_{ij} = \begin{cases} w_{ij} & \text{object } i \text{ has an affinity to object } j \text{ with magnitude } w_{ij} \\ 0 & \text{otherwise} \end{cases}$$

As before, we assume that affinities are symmetric. The degree also generali in the obvious way

$$d_i = \sum_{j=1}^{n} A_{ij}$$

The degree matrix of an adjacency matrix is a diagonal matrix, where diagonal entries are the (unweighted or weighted) degrees of the correspond objects

$$D_{ii} = d_i$$

If the rows of an adjacency matrix are divided by the (weighted) degr then the sum of each row is 1, and it is natural to interpret the entries defining a kind of probability associated with each edge of the graph. T matrix is called the *walk matrix*

$$W_{ij} = \begin{cases} w_{ij}/d_i & \text{object } i \text{ has an affinity to object } j \text{ with magnitude } w_{i} \\ 0 & \text{otherwise} \end{cases}$$

The walk matrix provides one intuition about the composition of affinit in terms of the properties of random walks on the graph. We interpret off-diagonal entries of row i as the *transition probabilities* of moving to r nodes from node i. If we consider two graph nodes, say a and b, then number of steps it takes a random walk starting from a to reach b is measure of the global affinity between them. The number of steps that random walk takes from a depends on the length of the path between the but also on how many other possible paths branch off along the way, lead to long, circuitous paths back to b. Hence such a random walk capture great deal of information about the 'geography' of the graph between a b. An important application of this idea is used by Google to generate ranking of web pages that is used in its search engine.

4.3   Eigenvalues and eigenvectors

Given a symmetric matrix A, of size $n \times n$, an eigenvalue-eigenvector (v, λ), where v is a vector of length n and λ is a scalar, satisfies

$$Av = \lambda v$$

The usual explanation of such pairs is that an eigenvector represents a vector that, when acted on by the matrix, doesn't change direction but change magnitude by a multiplicative factor, λ.

This explanation is not particularly helpful in a data-mining setting since it isn't particularly obvious why the *action* of a matrix on some vector space should reveal the *internal structure* of the matrix. A better intuition the following. Suppose that each node of the graph is allocated some value and this value flows along the edges of the graph in such a way that the outflowing value at each node is divided up in proportion to the magnitud of the weights on the edges. Even though the edges are symmetric, the globa distribution of value will change because the two nodes at each end of a edge usually have a different number of edges connecting to them. Differen patterns of weighted edges lead to the value accumulating in different amount at different nodes.

An eigenvector is a vector of size n and so associates a value with eac node of the graph. In particular, it associates a value with each node such tha another round of value flow doesn't change each node's relative situation. T put it another way, each eigenvector captures an invariant distribution of valu to nodes, and so describes an invariant property of the graph described by A The eigenvalue, of course, indicates how much the total value has changed but this is uninteresting except as a way of comparing the importance of on eigenvalue-eigenvector pair to another.

You may recall the *power method* of computing the principal eigenvec tor of a matrix: choose an arbitrary vector, and repeatedly multiply it b the matrix (usually scaling after each multiplication). If A is a weighted ad jacency matrix, then the entries of its, say, pth power describe the weight along paths of length p. If such a matrix has no net effect on a particular allo cation of values to the graph nodes, we can think of this as being because th values have been passed along a loop of length p that ends where it started The effectiveness of the power method shows that the principal eigenvector i related to long loops in the graph.

4.4 Connections to SVD

Although we avoided making the connection in the previous chapter, SVD is a form of eigendecomposition. If A is a rectangular matrix, then the it column of U is called a left singular vector, and satisfies

$$A'u_i = s_i v_i$$

and the ith column of V is called a right singular vector, and satisfies

$$Av_i = s_i u_i$$

In other words, the right and left singular vectors are related properties as ciated with, respectively, the attributes and objects of the dataset. The act of the matrix A is to map each of these properties to the other. In fact, singular value decomposition can be written as a sum in a way that ma this obvious

$$A = \sum_{i=1}^{m} s_i u_i v_i$$

The connection to the correlation matrices can be seen from the eq tions above, since

$$AA'u_i = s_i A v_i = s_i^2 u_i$$

so that u_1 is an eigenvector of AA' with eigenvalue s_i^2. Also

$$A'Av_i = s_i A'u_i = s_i^2 v_i$$

so v_i is an eigenvector of $A'A$ with eigenvalue s_i^2 also.

With this machinery in place, we can explore one of the great succe stories of eigendecomposition, the PageRank algorithm that Google uses rank pages on the world wide web. This algorithm shows how a graph structure can reveal properties of a large dataset that would not otherwise obvious.

4.5   Google's PageRank

An important application of these ideas is the PageRank algorithm t Google uses to rank web pages returned in response to search queries. S isfying a search query requires two different tasks to be done well. Fi pages that contain the search terms must be retrieved, usually via an inc Second, the retrieved pages must be presented in an order where the n significant pages are presented first [19–21]. This second property is par ularly important in web search, since there are often millions of pages t contain the search terms. In a traditional text repository, the order of sentation might be based on the frequencies of the search terms in each of retrieved documents. In the web, other factors can be used, particularly extra information provided by hyperlinks.

The pages on the web are linked to each other by hyperlinks. starting point for the PageRank algorithm is to assume that a page is lin to by others because they think that page is somehow useful, or of high qual In other words, a link to a page is a kind of vote of confidence in the importa of that page.

Suppose that each page on the web is initially allocated one unit importance, and each page then distributes its importance proportionally

all of the pages it points to via links. After one round of distribution, a
pages will have passed on their importance value, but most will also hav
received some importance value from the pages that point to them. Page
with only outgoing links will, of course, have no importance value left, sinc
no other pages point to them. More importantly, after one round, page
that have many links pointing to them will have accumulated a great deal c
importance value.

Now suppose we repeat the process of passing on importance value in
second round, using the same proportionate division as before. Those page
that are pointed to by pages that accumulated lots of importance in the firs
round do well, because they get lots of importance value from these upstrear
neighbors. Those pages that have few and/or unimportant pages pointing t
them do not get much importance value.

Does this process of passing on importance ever converge to a stead
state where every node's importance stays the same after further repetitions
If such a steady state exists, it looks like an eigenvector with respect to som
matrix that expresses the idea of passing on importance, which we haven
quite built yet. In fact, it will be the principal eigenvector, since the repeate
passing around of importance is expressed by powers of the matrix.

The matrix we need is exactly the walk matrix defined above, excep
that this matrix will not be symmetric, since a page i can link to a pag
j without there having to be a link from j to i. The recurrence describe
informally above is

$$x_{i+1} = x_i W$$

where W is the directed walk matrix, and x_i is the $1 \times n$ vector that describe
the importances associated with each web page after round i.

There are several technical problems with this simple idea. The firs
is that there will be some pages with links pointing to them, but no link
pointing from them. Such pages are sinks for importance and, if we kee
moving importance values around, such pages will eventually accumulate a
of it. The simple solution is to add entries to the matrix that model link
from such pages to *all* other pages in the web, with the weight on each lin
$1/n$, where n is the total number of pages indexed, currently around 8 billior
In other words, such sink pages redistribute their importance impartially t
every other web page, but only a tiny amount to each. We can think of th
as *teleportation* of importance, since it no longer flows along links, but jum
from one part of the graph to another.

This same problem can also occur in entire regions of the web; there ca
exist a subgraph of the web from which no links emanate, although there ar
links within the subgraph. For example, the web sites of smaller companie
may well contain a rich set of links that point to different parts of their we
site, but may not have any links that point to the outside web. It is likely tha

there are many such regions, and they are hard to find, so Google modi
the basic walk matrix to avoid the potential problem, rather than find
occurrences and dealing with them explicitly.

Instead of using W, Google uses W_{new}, given by

$$W_{new} = \alpha W + (1 - \alpha)E \qquad ($$

where E is an $n \times n$ matrix generalizes the idea of teleporting importance fr
a single node to a description of how importance teleports between every p
of nodes, and α is between 0 and 1 and specifies how much weight to alloc
to the hyperlinked structure of the web (the first term), and how much to
teleportation described by E (the second term).

The matrix E was originally created to avoid the problem of region
the graph from which importance could not flow out via links. However, it
also be used to create importance flows that can be set by Google. Web s
that Google judges not to be useful, for example web spam, can have th
importance downgraded by making it impossible for importance to telep
to them.

The result of this enormous calculation (n is of the order of 8 billi
is an eigenvector, whose entries represent the amount of importance t
has accumulated at each web page, both by traversing hyperlinks and
teleporting. These entries are used to rank all of the pages in the web
descending order of importance. This information is used to order sea
results before they are presented.

The surprising fact about the PageRank algorithm is that, althoug
returns the pages related to a query in their *global* importance rank order,
seems adequate for most searchers. It would be better, of course, to ret
pages in the importance order relevant to each particular query, which sc
other algorithms, notably HITS, do [69].

The creation of each updated page ranking requires computing the p
cipal (largest) eigenvector of an extremely large matrix. Some care must
taken in implementing the recurrence (4.1) because W_{new} is now a dense
trix because of E. However, a fast update is possible by rearranging the or
of the operations. Theory would suggest that it might take (effectively) a la
number of rounds to distribute importance values until they are stable.
actual algorithm appears to converge in about 100 iterations, so presuma
this is still some way from stability – but this may not matter much given
other sources of error in the whole process. The complexity of this algorit
is so large that it is run in a batched mode, so that it may take several d
for changes in the web to be reflected in page rankings.

PageRank is based on the assumption that links reflect opinions ab
importance. However, increasingly web pages do not create links to ot

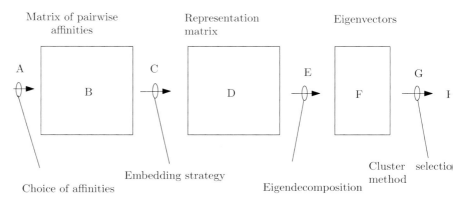

Figure 4.2. *The global structure of analysis of graph data.*

pages because it's easier and faster to find them again by searching at Google
It is clear that a new way to decide which pages are important needs to b
developed.

4.6 Overview of the embedding process

We now turn to ways in which we might discover structure, particularly clus
tering, within a graphical dataset. As we noted above, we could work directl
in the graph space, but the complexities of the algorithms often make th
impractical.

Instead, we find ways to embed the graph space in a geometric space
usually a Euclidean space, in a way that preserves relationships among object
appropriately. Figure 4.2 provides an overview of the entire process.

Here is a brief description of the phases:

- Arrow A describes an initial transformation from a Euclidean space int
  a graph or affinity space. Although this seems to be a retrograde step,
  can be appropriate when the data in the Euclidean space cannot easil
  be clustered directly. This is usually because the clusters are highl
  non-convex, or because the data occupies a low-dimensional manifold i
  a high-dimensional space.

- Matrix B is an $n \times n$ matrix of affinities, positive values for which a larg
  magnitude indicates a stronger affinity. The matrix is usually, thoug
  not always, symmetric, that is the edges in the graph are considered t
  be undirected.

- Arrow C is the critical step in the entire process. It maps the affin matrix to a representation matrix in such a way that geometric relati ships in the geometric space reflect non-local relationships in the gra space. As a result, good clusterings in the geometric space are also go clusterings in the graph space.

- This produces matrix D which behaves like a dataset matrix in previous chapter; that is it describes points in a geometric space wh distances apart match, as much as possible, sensible separations in original graph. However, matrix D is still $n \times n$, so its distances are well-behaved, and it is hard to work with.

- Arrow E represents the mapping of the geometric space described D to a lower-dimensional space where distances are better behav and so clustering is easier. We can use the techniques described in previous chapter, except that matrix D is square so we can use eigen composition instead of SVD if we wish. The mapping of affinity mat to representation matrix can also sometimes provide extra informat about the reduction to a lower-dimensional space.

- Matrix F has columns consisting of the k most significant eigenvect of D, so F is an $n \times k$ matrix. It corresponds to the U matrix in SVD. The rows of F can be regarded as defining coordinates for e object (node) in a lower-dimensional space.

- Finally, arrow G is the process of clustering. We have discussed ma of these clustering techniques in the previous chapter.

4.7   Datasets versus graphs

It is hard to visualize a graph space, and to see how it differs from a geome space. We can get some intuition for the differences using a small exampl

Consider the affinity matrix:

$$\begin{bmatrix} 0 & 1 & 1 & 1 & 1 \\ 1 & 0 & 0 & 0 & 1 \\ 1 & 0 & 0 & 1 & 0 \\ 1 & 0 & 1 & 0 & 0 \\ 1 & 1 & 0 & 0 & 0 \end{bmatrix}$$

If we think of this as an ordinary dataset, and the rows as points in a fi dimensional space, then the first row corresponds to a point that is far fr the origin, while the other rows are all closer to the origin. If we cluster directly, the object corresponding to row 1 forms a cluster by itself far fr the others, objects 2 and 3 form a cluster and objects 4 and 5 form anot cluster, although these last two clusters are not well-separated, as you might expect.

However, from a graph point of view, the object corresponding to row is the center of the graph and connects the rest of the graph together because of its links to all of the other objects. A clustering with respect to the graph structure should place object 1 centrally, and then connect the other objects building from that starting place.

The significance and placement suggested by the geometric space view is exactly 'inside out' from the significance and placement suggested by the graph space view – large values that place an object far from the origin, and so far from other objects, in the geometric space correspond to tight bonds that link an object closely to other objects in the graph space.

If we want to embed a graph space in a geometric space in a way that makes graphical properties such as centrality turn out correctly, we are going to have to include this inside-out transformation as part of the embedding and indeed we will see this happening in representation matrices.

Embedding using adjacency matrices was popular a decade or more ago but more recently embeddings based on Laplacians and their relatives are used precisely because of this need to turn graph structures inside out before embedding them.

4.7.1   Mapping Euclidean space to an affinity matrix

The first possible step, as explained earlier, is not necessarily common in data mining, except in a few specialized situations. Sometimes, a dataset appears extremely high-dimensional but it is known from the problem domain that the 'real' dimensionality is much lower. The data objects actually lie on low-dimensional manifold within the high-dimensional space, although the manifold may have a complex, interlocking shape. For example, a complex molecule such as a protein can be described by the positions of each of its atoms in three-dimensional space, but these positions are not independent, so there are many fewer degrees of freedom than there appear to be. It may also be that the placing of the objects makes it hard for clustering algorithms with built-in assumptions about convexity of clusters, to correctly determine the cluster boundaries.

In such settings, it may be more effective to map the dataset to an affinity matrix, capturing local closeness, rather than trying to reduce the dimensionality directly, for example by using an SVD. When this works, the affinity matrix describes the local relationships in the low-dimensional manifold, which can then be unrolled by the embedding.

Two ways to connect objects in the high-dimensional dataset have been suggested [12]:

1. Choose a small value, ε, and connect objects i and j if their Euclid
   distance is less than ε. This creates a symmetric matrix. However, i
   not clear how to choose ε; if it is too small, the manifold may diss
   into disconnected pieces.

2. Choose a small integer, k, and connect object i to object j if object
   within the k nearest neighbors of i. This relationship is not symmet
   so the matrix will not be either. Because k describes the cardinality
   set of neighbors, it is much less sensitive to the distribution of distan
   at small scales.

There are also different possible choices of the weight to associate with e
of the connections between objects. Some possibilities are:

1. Use a weight of 1 whenever two objects are connected, 0 otherwise; t
   is A is just the adjacency matrix induced by the connections.

2. Use a weight

$$A_{ij} = exp(-d(x_i, x_j)^2/t)$$

   where $d(x_i, x_j)$ is a distance function, say Euclidean distance, and t
   scale parameter that expresses how quickly the influence of x_i spre
   to the objects near it. This choice of weight is suggested by connect
   to the heat map [12].

3. Use a weight

$$A_{ij} = exp(-d(x_i, x_j)^2/t_1 t_2)$$

   where t_i and t_j are scale parameters that capture locally varying den
   of objects. Zelnik-Manor and Perona [117] suggest using $t_i = d(x_i,$
   where x_k is the kth nearest neighbor of x_i (and they suggest $k = 7$).
   when points are dense, t_i will be small, but when they are sparse t_i
   become larger.

These strategies attempt to ensure that objects are joined in the graph sp
only if they lie close together in the geometric space, so that the structure
the local manifold is captured as closely as possible, without either omitt
important relationships or connecting parts of the manifold that are logica
far apart.

4.7.2   Mapping an affinity matrix to a representation mat

The mapping of affinities into a representation matrix is the heart of an
fective decomposition of the dataset.

The literature is confusing about the order in which eigenvalues are being talked about, so we will adopt the convention that eigenvalues (like singular values) are always written in descending order. The largest and the first eigenvalue mean the same thing, and the smallest and the last eigenvalue also mean the same thing.

Representation matrix is the adjacency matrix

We have already discussed the most trivial mapping, in which the representation matrix is the adjacency matrix of the affinity graph. The problem with this representation is that it fails to make the inside-out transformation discussed earlier, and so starts from an inaccurate representation to produce clustering of the graph.

Historically, adjacency matrices have been studied because they are easy to understand, and can reveal some of the properties of a graph. For example the eigenvalues of the adjacency matrix have this property

$$\text{minimum degree} \leq \text{largest eigenvalue} \leq \text{maximum degree}$$

Global graph properties such as betweenness and centrality that are of interest in *social network analysis* can also be calculated from the adjacency matrix.

Representation matrix is the walk matrix

There are two possible normalizations of the adjacency matrix. The first we have already seen, the walk matrix that is obtained from the adjacency matrix by dividing the entries in each row by the sum of that row. In matrix terms $W = D^{-1}A$. This matrix can be interpreted as the transition probabilities of a random walk, and we saw how this can be exploited by the PageRank algorithm. Again this matrix is not a good basis for clustering, but some global properties can be observed from it.

Representation matrix is the normalized adjacency matrix

The second normalization of the adjacency matrix is the matrix

$$N = D^{-1/2}AD^{-1/2}$$

where D is the matrix whose diagonal entries are the reciprocals of the square roots of the degrees, i.e. $1/\sqrt{d_i}$. This matrix is symmetric.

The matrices W and N have the same eigenvalues and a one-to-one relationship between their eigenvectors: if w is an eigenvector of W then there is eigenvector, v, of N, such that $w = vD^{1/2}$.

Representation matrix is the graph Laplacian

It turns out that the right starting point for a representation matrix t
correctly captures the inside-out transformation from graphs to geome
space is the *Laplacian* matrix of the graph. Given pairwise affinities,
matrix is

$$L = D - A$$

that is L is the matrix whose diagonal contains the degrees of each node, a
whose off-diagonal entries are the negated values of the adjacency matrix

At first sight, this is an odd-looking matrix. To see where it cor
from, consider the *incidence matrix* which has one row for each object,
column for each edge, and two non-zero entries in each column: a $+1$ in
row corresponding to one end of the edge, and a -1 in the row correspond
to the other (it doesn't matter which way round we do this for what a
underneath, undirected edges). Now the Laplacian matrix is the prod
of the incidence matrix with its transpose, so the Laplacian can be thou
of as a correlation matrix for the objects, but with correlation based o
on connection. Note that the incidence matrix will, in general, have m
columns than rows, so this is one example where the correlation matri
smaller than the base matrix.

L is symmetric and positive semi-definite, so it has n real-valued eig
values. The smallest eigenvalue is 0 and the corresponding eigenvector is
vector of all 1s. The number of eigenvalues that are 0s corresponds to
number of connected components in the graph. In a data-mining setting,
means that we may be able to get some hints about the number of clust
present in the data, although connected components of the graph are e
clusters.

The second-smallest eigenvalue-eigenvector pair is the most interest
in the eigendecomposition of the Laplacian. If the eigenvalue is non-zero t
the graph is connected. The second eigenvector maps the objects, the no
of the graph, to the real line. It does this in such a way that, if we choose a
value, q, in the range of the mapping, and consider only those objects map
to a value greater than or equal to q, then those objects are connected in
graph. In other words, this mapping arranges the objects along a line in a
that corresponds to sweeping across the graph from one 'end' to the ot
Obviously, if we want to cluster the objects, the ability to arrange them
this way is a big help. Recall that a similar property held for the first colu
of the U matrix in the previous chapter and we were able to use this to clus
objects.

In fact, all of the eigenvectors can be thought of as describing mo
of vibration of the graph, if it is thought of as a set of nodes connected
slightly elastic edges. The second smallest eigenvector corresponds to a m

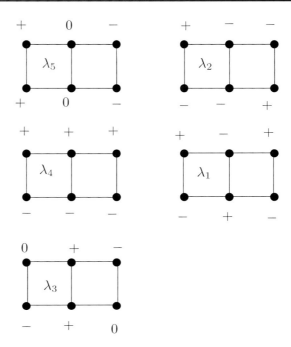

Figure 4.3. *Vibration modes of a simple graph.*

in which, like a guitar string, half the graph is up and the other half is down
The next smallest eigenvector corresponds to a mode in which the first and
third 'quarters' are up, and the second and fourth 'quarters' down, and so on
This is not just a metaphor; the reason that the matrix is called a Laplacian is
that it is a discrete version of the Laplace-Beltrami operator in a continuou
space.

This pattern of vibration can be seen in the small graph shown in Fig
ure 4.3. The top left graph shows the mode corresponding to the second small
est eigenvalue, with '+' showing where the eigenvector values are positive, '0
showing where the eigenvector is 0, and '−' showing where the eigenvectc
values are negative. As expected, this mode shows eigenvector values increas
ing from one 'end' of the graph to the other. Directly below this is the grap
showing the mode corresponding to the third smallest eigenvalue. Now th
values increase 'across' the graph. The graph at the bottom of the first co
umn shows the mode of the next smallest eigenvalue which is a more comple
pattern; the two graphs in the second column correspond to the second large
and largest eigenvalues, and show even more complex patterns. In genera
the larger the eigenvalue, the smaller the connected regions with positive an
negative eigenvector values, and the more such regions into which the grap
is divided.

Obviously, the eigenvectors corresponding to small eigenvalues prov information about where important boundaries in the graph lie – after all, edges between regions that are up and down in a vibration mode are pla where the graph can 'bend' in an interesting way, and these correspond boundaries between good clusters.

Another property of the Laplacian matrix, L, is that for every vecto of n elements

$$v'Lv = \frac{1}{2} \sum_{i,j=1}^{n} w_{ij}(v_i - v_j)^2$$

where w_{ij} are the affinities, and v_i and v_j are elements of v. This equatio useful for proving many other properties of the Laplacian, and also expla the connection of Laplacians to quadratic minimization problems relating cuts.

Partitioning the geometric space whose representation matrix is Laplacian corresponds to minimizing the ratio cut in the graph. The re cut of a graph G divided into a subset S and its complement $\overline{S}$ is

$$\phi(S) = \frac{|E(S, \overline{S})|}{min(|S|, |\overline{S}|)}$$

where E is the number of edges between the two subsets separated by the c Define

$$\Phi(G) = min\ \phi(S)$$

over all possible cuts. $\Phi(G)$ is called the *isoperimetric number* of the gra and has many connections to other properties of the graph. For examp Cheeger's inequality says that

$$\Phi(G) \leq \lambda_{n-1} \leq \frac{\Phi(G)}{2d}$$

where d is the maximum degree of the graph.

Representation matrix is the walk Laplacian

There are two ways to normalize the Laplacian matrix, analogous to the t ways to normalize an adjacency matrix. The first is the walk Laplacian, given by

$$L_w = D^{-1}L = I - D^{-1}A$$

where the entries in each row are divided by the (weighted) degree. Note second equality which shows explicitly how this matrix turns the adjace matrix inside out.

Partitioning the geometric space given by the walk Laplacian corr
sponds to minimizing the normalized cut of the graph. The *normalized c
(NCut)* of a graph G divided into a subset S and its complement $\overline{S}$ is

$$NCut(S) = \frac{|E(S, \overline{S})|}{vol(S)}$$

where $vol(S)$ is the total weight of the edges in S. The NCut is small when
cut is balanced both in terms of nodes and edges.

This embedding has the property that nodes are placed closer togeth
when there are both short paths and multiple paths between them in th
graph. There seems to be a consensus that this matrix, and the resultir
embedding, is the most appropriate for most datasets.

Representation matrix is the normalized Laplacian

The second way to normalize the Laplacian gives the normalized Laplacian

$$L_n = D^{-1/2}LD^{-1/2} = I - D^{-1/2}AD^{-1/2}$$

L_n and L_w are related: they have the same eigenvalues and relate
eigenvectors. An eigenvector v of L_n satisfies $L_n v = \lambda v$ while a (generalizec
eigenvector v of L_w satisfies $L_w v = \lambda D v$.

This embedding is much the same as the previous one, but less numer
cally stable.

Representation matrix is the pseudoinverse of the Laplacian

Another way of thinking about distances in the graph is to consider how lon
it takes a random walk, with transition probabilities weighted by the weigh
on the edges, to go from a source node to a destination node. This natural
treats nodes with many paths between them as closer because there are mon
ways for the random walk to get between them. We can define the *hittir
time*

$$h(i \rightarrow j) = \text{average number of steps to reach } j \text{ from } i$$

This measure, however, is not symmetric, that is $h(i \rightarrow j) \neq h(j \rightarrow i)$, so
turns out to be more convenient to define the *commute time*

$$c(i, j) = h(i \rightarrow j) + h(j \rightarrow i)$$

and this is symmetric. The commute time measures the average time for
random walk on the weighted graph to leave node i and return to it, havin

passed through node j. The commute time measures both how close no
i and j are along a 'direct' path between them, but also how many possi
detours leading to much longer paths between them are encountered al
the way.

The commute times and their square roots from the graph behave
Euclidean distances, so we would like to be able to create a representat
matrix containing them. This would be extremely difficult to compute dired
– even a single commute distance is not an easy measure to compute. Bu
turns out that commute distance can be computed using the Moore-Penr
pseudoinverse of the Laplacian of the graph, which we write as L^+.

L is not of full rank since, even if the graph is connected, L has one z
eigenvalue. Hence L does not have an inverse. The pseudoinverse of a mat
behaves much like an inverse in most contexts, and exists for any mat
even one that is not square or does not have full rank. The properties of
pseudoinverse are

$$L L^+ L = L$$
$$L^+ L L^+ = L^+$$
$$(LL^+)' = LL^+$$
$$(L^+L)' = L^+L$$

If L^+ has entries l_{ij}^+, and $vol(G)$ is the total number of edges in
graph then the commute distance between nodes i and j is

$$c(i,j) = 2\,vol(G)\,(l_{ii}^+ + l_{jj}^+ - 2l_{ij}^+)$$

There is a strong connection to electrical networks – in fact the right-ha
term in parentheses is the effective resistance between the two points i an
if the edges of the graph are regarded as wires whose resistance is invers
proportional to their pairwise affinities.

Once we know the pseudoinverse of the Laplacian, then we can trivia
build an $n \times n$ representation matrix whose entries are the square roots
commute distances, and this matrix embeds the graph into Euclidean spa

Unfortunately, computing the pseudoinverse is non-trivial for large gra
Two methods, both of which avoid computing the entire matrix, have b
suggested by Fouss $et\ al.$ [43], and Brand [16].

4.8   Eigendecompositions

The representation matrix is a square $n \times n$ matrix, but is still difficult to wo
with because it is sparse, since it reflects to some extent, the sparseness of

adjacency matrix. It is also high-dimensional, since n objects are placed in space of dimension n. It is natural to use the techniques from the previou chapter to reduce the dimensionality, especially as it is likely that the manifol described by the embedded geometric space has much smaller dimensionalit than it appears to have – the points could have been embedded at the vertice of a high-dimensional tetrahedron, but it is not likely. Therefore, we expec that large-scale dimension reduction is possible.

Instead of using SVD, we can use the slightly simpler eigendecompos tion, since the representation matrix, R, is square. The eigendecompositio expresses R as the product

$$R = P \Lambda P^{-1}$$

where P is a matrix whose columns are the (orthogonal) eigenvectors, and is a diagonal matrix whose entries are the eigenvalues. Recall we are assumin that the eigenvalues are presented in decreasing order just as in SVD (not a software implementations of eigendecomposition will do this).

Just as with SVD, we can examine the magnitudes of the eigenvalue and use them to choose how many columns of P to retain, say k. The row of the truncated P matrix can be treated as coordinates in a k-dimensiona space.

For the eigendecomposition starting from the Laplacian and the wal Laplacian, the eigenvectors are indicator vectors; that is their positive an negative entries divide the objects into two subsets, as long as the magnitud of the entries are bounded away from zero. Small magnitude entries are prot lematic, since eigendecompositions are robust under perturbations, and so value close to zero could possibly be on the 'wrong' side. One of the weak nesses of the normalized Laplacian is that its eigenvectors are not necessaril indicator vectors; in particular, nodes of unusually low degree are hard t classify [115].

It is also necessary that the order of the eigenvectors is significant c else we may lose clusters when we truncate at some k. For representatio matrices derived from the Laplacian, the order of eigenvectors is significan but not necessarily so for representation matrices derived from the adjacenc matrix. Clustering based on the adjacency matrix will work properly if ther are indeed strong clusters in the data, but may fail to work when clusters ar hard to separate. Such clusterings are harder to justify.

4.9 Clustering

The truncated version of the representation matrix, obtained via the eigend composition, is a faithful low-dimensional representation of the matrix. W can therefore cluster the objects using techniques that we have seen befor

However, there are several ways in which we can use the fact that the d
came from a graph space.

Using the Fiedler vector

The eigenvector corresponding to the second smallest eigenvalue, λ_{n-1}, of
Laplacian, is called the Fiedler vector. Recall that if we sort the values in
eigenvector into ascending order and separate the rows (objects) by whet
they are greater than or equal to; or less than or equal to some chosen val
then the nodes in each separated group are connected. This tells us tha
simple way to divide the graph is to choose a value near the middle of
range, typically zero, and use that to split the graph into two equal-si
pieces. Of course, the fact that this gives rise to connected clusters in
graph does not necessarily mean that these are good clusters. By the sa
logic, we can choose three equally spaced values and divide the graph into
pieces and so on, although we are not guaranteed that nodes in the mic
intervals are connected.

Simple clustering

A second approach that has often been suggested in the literature is to
some standard clustering algorithm, such as k-means to cluster the data ba
on its coordinates in the k-dimensional space. The idea is that the geome
space has been reduced to its essentials, with noise removed and the d
expressed in its lowest-dimensional terms. Therefore, even a simple cluster
technique should be effective.

Clustering using eigenvectors directly

Clustering using the Fiedler vector relies on the structure captured by
eigenvector of the $(n-1)$st (second smallest) eigenvalue. In general, us
more of the eigenvectors will produce a better clustering.

The truncated matrix places a point (or vector) for each object in a
dimensional space. The geometry of this space can be exploited in a num
of ways. Alpert and Yao [7] provide a summary of some possible approach

Clustering on the unit sphere

The geometric space into which a graph space has been embedded can inh
some structure from it. For example, if the graph contains a central n
that is well-connected to much of the rest of the graph, then this node will

placed close to the origin because it is being pulled by many other nodes, bu these other nodes in turn will be pulled towards the center.

Another way to think about this is that the stationary structure is sen sitive to the (weighted) degrees of the nodes in the graph. Nodes with hig degree have many immediate neighbors, but typically also many neighbo: slightly further away. In some settings, this can distort the modelling goa When the goal is to find the nearest interesting neighbor or to predict a ne graph edge, the presence of such a node can make the problem much harde Its presence overwhelms the remaining structure of the graph.

For example, in collaborative filtering it is not obvious how to treat neutral opinion. If an individual sees a movie and neither likes not dislikes i this does not provide any more objective information about how to recommen the movie to others than if that individual had not rated it at all. However, tl addition of that edge to the graph of recommendations provides potential' many paths between *other* pairs of nodes, which now seem closer, even thoug nothing substantive has been added to the body of knowledge. It is useful t be able to factor out the kind of generic popularity that is reflected in suc neutral ratings from more-useful rating information. (Neutral ratings at not necessarily completely useless – a movie that has been neither liked nc disliked by many people is a mediocre movie; one that has hardly been rate at all is of unknown quality. It seems difficult to reflect this kind of secon order information well in collaborative filtering. This difficulty is visible, fc example, in Pandora, the music recommendation system.)

When the geometric space has been derived from the pseudoinverse the Laplacian, based on commute distances, then the distance of any obje from the origin is $\sqrt{l_{ii}}$, which can be interpreted as the reciprocal of the gener popularity of this state. The effect of this generic popularity can be remove by removing the effect of distance from the origin as part of the distant measure between objects. If we map each object to the surface of the un sphere in k dimensions, then we can use the cosine distance for clusterin, based on a more intrinsic similarity between objects.

The non-local affinity based on angular distance is

$$\cos(i,j) = l_{ij}^+ / \sqrt{l_{ii}^+ l_{jj}^+}$$

which normalizes in a way that discounts generic popularity.

Examples

Shi and Malik [99] and Belkin and Niyogi [12] both use the unnormalize Laplacian, and find solutions to the generalized eigenvalue problem Lv

λDv. Shi and Malik then suggest using k-means to cluster the points in resulting k-dimensional space.

Ng, Jordan and Weiss [89] use the normalized Laplacian, compute $n \times k$ matrix of eigenvectors, and then normalize the rows to have norm This addresses the problem of low-degree nodes. They then use k-means cluster the points.

Meilă and Shi [86] use the walk adjacency matrix, and then cluster rows of the $n \times k$ matrix of eigenvectors.

It is not clear who deserves the credit for suggesting that the best r resentation matrix is the walk Laplacian, although some of the arguments its favor can be found in [114] and a useful discussion in [95].

4.10 Edge prediction

Given a graph, especially one that describes a social network, it may be teresting to predict the existence of one or more edges or links in the gra that are not present. For example, if the graph is dynamic and grows c time, such a prediction indicates an edge that is likely to appear in the futi perhaps the most likely to appear. If the graph describes the interactions (group of terrorists or criminals, such a link may indicate an interaction t is present in reality but has not been observed. In a more mundane setti in a graph of streets weighted by the traffic they carry, an edge predict indicates a possible new traffic route that would be well-used if built.

The intuition behind edge prediction is that two nodes should be joi if they are *close* in the graph. What closeness means, as we have seen, vary quite a lot. In its simplest form, the two closest nodes might be th with the shortest (weighted) path between them. However, this is a quit weak measure of nearness.

Liben-Nowell and Kleinberg [81] have experimented with predicting ed using a wide variety of closeness measures, some of them based on propert derived from the two nodes' mutual neighbors. These range from simple m sures such as how many neighbors they have in common, to more comp measures such as their Jaccard coefficient (the ratio of the number of comn neighbors they have to the total number of neighbors they have). They a consider measures that take into account the entire structure of the gra including the commute time modulated to reduce the effects of stationar and a measure defined by Katz which takes the form $(I - \beta A)^{-1} - I$ (where is the adjacency matrix). This score depends on the number of paths betw each pair of nodes, with β a scaling parameter that determines the relat weight of long versus short paths. Liben-Nowell and Kleinberg experim with a number of datasets, and overall they achieve prediction accuracies

the order of 10–15%, about 30 times better than chance. We might expec
that the more sophisticated measures of nearness used to drive the embec
dings described in this chapter might do better, and there is clearly plenty c
room for improvement.

4.11   Graph substructures

In some contexts, it may be useful to look at graph structure that is nc
necessarily of global interest. Many law enforcement and intelligence dat
gathering leads to graph data structures because it is the *relationships* betwee
entities, rather than properties of the entities themselves, that is of primar
interest. For example, the U.K. police use a system called HOLMES 2 (Hom
Office Large Major Enquiry System)[2] [84, 112] that is used in an investigatio
to record all individuals, addresses, actions, statements, and descriptions, a
well as the results of surveillance of, for example, credit card usage. Graph
connecting these various entities can be displayed and traversed.

Normal relationships in large-scale graph data should appear many time
so the structures corresponding to them should be common. Structures tha
are not common may represent relationships that are abnormal, and so c
interest. The difficulty is that typical graphs are large and contain many di
ferent substructures, so it is hard to discover places where something unusu;
is going on. Even visualization, which itself requires sophisticated graph
drawing algorithms, is of limited usefulness.

One approach to this problem is to construct patterns (subgraphs) tha
correspond to known activities of interest. For example, drug dealing ma
result in a web of connections joining a cash business to a number of ir
dividuals who have been detected carrying amounts just under $10,000 t
particular bank accounts. Discovering drug dealers becomes a kind of graph
ical information-retrieval problem, looking inside large graphs for particula
substructures. Unfortunately, if not all edges are present in the graph, pe
haps because some relationship was not observed, then the pattern match wi
fail. This approach also depends on understanding all possible relationship
that correspond to activities of interest – a clever criminal may invent a ne
strategy that will not be detected by searching for known patterns.

Ways to examine graphs that find *all* unusual structures avoid thes
weaknesses. The set of unusual structures at least contains the set of activitie
of interest, although it might contain other activities as well. It will not, i
general, contain structures corresponding to innocuous activities, since thes
will occur many times and so will not appear unusual.

The eigenvectors corresponding to the smallest eigenvalues of the repr
sentation matrix reveal vibrational modes where large, well-connected regior

---

[2]A good example of a *bacronym*, a name carefully chosen to produce a good acronym

move in the same direction (that is, have positive or negative eigenvector v
ues of similar magnitude). As we have seen, the eigenvectors of the last f
eigenvalues are good clusterings of the nodes because the boundaries betwe
vibrational regions are good cuts.

Choosing slightly larger-valued eigenvalues finds regions of the gra
that are locally well-connected, even if they are small. Such regions are go
clusters except that they are small in size, so that the partitions they repres
are unbalanced cuts of the graph. Such regions are still of interest, sir
they represent groups of nodes with unusual affinity at medium scale.
relationship graphs, ordinary people tend to have local relationships that
sufficient to connect them quickly to the rest of the graph (the 'six degrees
separation' phenomenon), while groups that are trying, for whatever reas
to hide their existence and purpose may be less well connected. Small regi
that are internally well-connected but poorly connected to the rest of
graph are prospects for further analysis.

If we forget that the representation matrix arises from the embedd
of a graph, then we can also consider the eigenvectors corresponding to
large eigenvalues. Those nodes that have large values from these eigenvect
(either positive or negative) are 'interesting' in the sense that we discussed
the previous chapter – they have unusual correlation, considered as objec
with other nodes. These nodes are those that have unusual neighborhoo
most obviously because they have unusually high degree. However, no
whose neighborhoods are unusual for other reasons will also be selected.

There is, however, a third and more interesting case. The eigenvect
associated with eigenvalues of medium size select regions of the graph rep
senting small, unusual structures in the graph. Such structures may represe
groups whose connections to one another are unusual, or patterns that m
represent anomalous activity.

If we decompose the representation matrix using SVD, the columns
the resulting U matrix are the eigenvectors in which we are interested. T
eigenvectors that reveal global clustering in the graph are the last few colum
of the U matrix (those that correspond to the non-zero eigenvalues); 1
eigenvectors that reveal unusual local neighborhood structure are the fi
few columns of the U matrix; and the eigenvectors that reveal unusual gra
substructure are near the 'middle' of the matrix. However, finding th
'middle' columns requires some effort.

We can estimate the 'vibrational energy' associated with each eigenv
tor by computing the mean of the absolute values of its entries. Eigenvect
for which this measure is large correspond to vibrations that are large,
cause the magnitude of the vibrations are large, or because many nodes
involved, or both. On the other hand, eigenvectors for which this value
small correspond to vibrations that are small, and such vibrations are lik

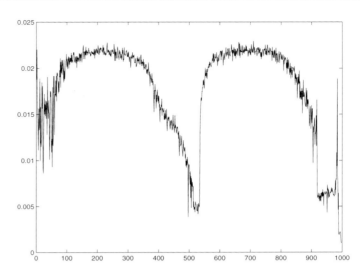

Figure 4.4. *Plot of the means of the absolute values of each colum of U. Low values indicate eigenvectors with low vibrational energy, eithe because many nodes move only slightly or because only a few nodes move Such components are likely to deserve further analysis.*

to correspond to interesting substructures. Hence we expect that this valu will be small for the first few and last few columns of U, and also for som columns close the to the middle of U.

We illustrate using a dataset of movie ratings, selected from the Movie Lens data [87], with 400 people rating 600 movies from 1 to 5, where 5 in dicates strong approval and 1 indicates strong disapproval. A graph is buil from this data by considering each individual and movie to be an object, an each non-zero rating to be an edge, producing a 1000×1000 matrix.

Figure 4.4 plots the means of the absolute values of the U matrix, ob tained by decomposing the walk Laplacian of this graph. It is obvious tha interesting structure is to be found associated with the eigenvectors at eac end, and a set of eigenvectors in the middle of the U matrix.

Figures 4.5–4.9 show a plot of a particular eigenvector, and the grap plotted using, as coordinates, that and an adjacent eigenvector. Hence prox imity in the graph plot corresponds to similarity according to these two eiger vectors. True graph edges are overlaid. We use eigenvectors from the regio of unusual node neighborhoods (50), an uninteresting region (250), the regio of unusual local substructure (500), another uninteresting region (750), and region of large clusters (910).

There are clear differences between the substructures associated wit

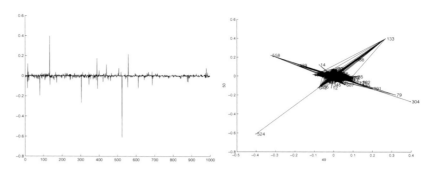

Figure 4.5.   *Eigenvector and graph plots for column 50 of the matrix . (See also Color Figure 1 in the insert following page 138.)*

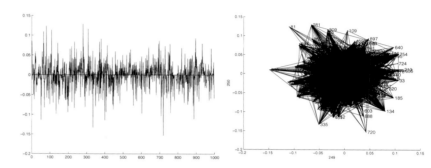

Figure 4.6.   *Eigenvector and graph plots for column 250 of the matrix. (See also Color Figure 2 in the insert following page 138.)*

components 250 and 750, which are dense and whose subgraphs contain m
overlapping structure, and the other components that involve relatively
nodes and whose subgraphs contain much simpler structures. The plot fr
component 50 indicates interesting single nodes; the plot from compon
910 indicates an interesting set of cliques (note how the extremal nodes
connected to each other); while the plot from component 500 indicates a m
complex structure involving relatively few nodes.

4.12   The ATHENS system for novel-knowledge discovery

A great deal of information is available in the web, but existing tools prov
only limited ways to find it. Conventional search engines, such as Google
Yahoo, are good at helping us to find out *more* about some topic, but
must know some relevant keywords first. Other systems like Vivisimo h

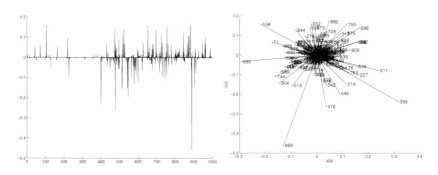

Figure 4.7. *Eigenvector and graph plots for column 500 of the* **(**matrix. *(See also Color Figure 3 in the insert following page 138.)*

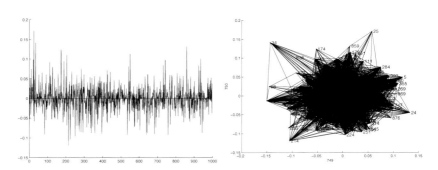

Figure 4.8. *Eigenvector and graph plots for column 750 of the* **(**matrix. *(See also Color Figure 4 in the insert following page 138.)*

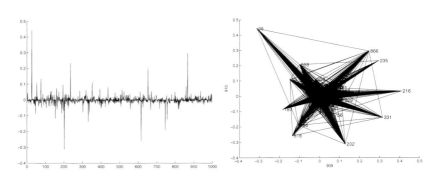

Figure 4.9. *Eigenvector and graph plots for column 910 of the* **(**matrix. *(See also Color Figure 5 in the insert following page 138.)*

to organize the results of a search into different groups, usually because different possible meanings of the search terms.

However, if we don't know about the existence of some topic, then th technologies are unable to help. ATHENS is a tool for discovering topics t: are new to us, not by randomly selecting topics (which would be another int esting approach), but by finding the most *relevant* topics that we do not kr about. In other words, ATHENS tries to impose some directionality in relationships among content pages in the web, and tries to search 'outwar from whatever a user considers a base.

The ATHENS tool can be used for individuals to discover what tl should learn next, that is topics for which they are well-prepared; for orga zations to discover what their next strategic step should be, or what directi their competitors are not well-prepared to go in; or for intelligence analy to avoid being trapped in silos of expectation.

The ATHENS tool has two major algorithmic components. The f is called *closure*. Given a set of pages, a closure aims to discover a robu but brief, description of the content of the pages. It does this by extract the most frequent words from the pages; comparing the frequencies in the of documents to the natural frequencies of the same words in English; a selecting those words that are more frequent in the returned documents. T set of words becomes a description of the content.

The second algorithmic component is called *probe*. Given two sets terms, a probe chooses pairs of terms, one from each set, and generate query to a search engine using the pair as search terms.

The ATHENS system is initially given a set of terms describing cont that is already understood, that is the user's model of the content in sc area of knowledge. To avoid too strong a dependence on the user's init choice of descriptive terms, these terms are used as a search query, a set appropriate pages returned, and a closure computed. This closure is desig: to be a good representation of what the user is aware of knowing.

Probes are then generated using the two lists: the initial set of ter from the user, and the description of the content returned by the closure. T returns a large collection of pages that are designed to be a good representat of what the user almost certainly knows, given the knowledge in the init domain. These pages do not yet represent novel knowledge.

A closure is then applied to the set of returned pages, primarily construct a master list of relevant words, and an association of these wo with each of the pages. An affinity matrix for the pages is constructed weighting an edge between each pair of pages based on the number of wo from the master list they share. A walk Laplacian is constructed from t affinity matrix and a clustering algorithm used to cluster the pages. A v short (three word) descriptor is generated for each cluster.

New probes are generated using the list of words from the primary clo
sure, and the short lists associated with each cluster. This is designed t
return knowledge that is new to the user (since such pages will typically no
contain *any* of the terms given by the user) but contextualized by what th
user probably already knows. Hence the retrieved set of pages represent
something 'just beyond' what the user knows.

Each probe generates a set of returned pages, that is there is one se
based on each of the clusters at the previous stage. Each set of returne
pages is processed as before: a closure is applied to the set to capture word
describing their content; these words are used to construct an affinity matri
and then a walk Laplacian; and the resulting graph is partitioned to produc
a set of clusters. These clusters represent novel knowledge and their conter
is described by a set of keywords (and, as the system is developed, by mor
sophisticated descriptors).

The ATHENS system has been applied to competitive intelligence [113
and to counterterrorism [104]. Further technical details can be found in [103

4.13   Bipartite graphs

Bipartite graphs are those in which the nodes can be divided into two classe
such that every edge passes from one class of nodes to the other. We hav
already seen that one way to interpret rectangular matrices is as bipartit
graphs, with one class describing the objects and the other class the attribute
This graph model of a dataset could be seen as a very simple way of mappin
from a high-dimensional geometric space to a graph space.

Bipartite graphs create a problem for the process we have outlined. Re
call that the graph of the web had the potential problem that value coul
become trapped in some subgraph. Bipartite graphs have the opposite prob
lem: the allocation of value never reaches a fixed point because value oscillate
from nodes of one kind to nodes of the other, and then back again. This intu
itively visible problem causes many technical problems with the embedding
we have been discussing.

One possible solution is to take the rectangular graph matrix for th
bipartite graph, say $n \times m$, and embed it in an $(n + m) \times (n + m)$ matri
as shown in Figure 4.10. If $n \gg m$, then the resulting matrix is not to
much larger. This is what we did with the movie rating data to create
graph in which we could look for unusual substructures. One advantage
this approach to handling bipartite graphs is that it also allows extensior
that relate objects of the same kind to each other, when this makes sens
(that is, the graph can be not quite bipartite). For example, Hendrickso
[52] suggests that word-document matrices, handled in this way, allow querie
involving both words and documents to be expressed simultaneously.

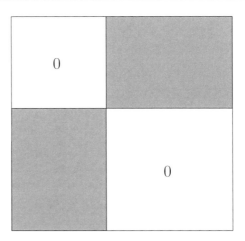

Figure 4.10. *Embedding a rectangular graph matrix into a square matr*

Notes

There is a large literature on the relationships between eigenvalues and eig
vectors of graphs, and other graph properties, for example [27, 28, 108]. So
of these global properties are of interest in data mining for what they tell
about a dataset. There is also a large literature in *social network anal*
where the graphs connecting individuals using many different kinds of aff
ties have been studied. Often the computation of appropriate properties
social network analysis has been done directly, which has limited the kinds
graphs that can be studied. This is changing with the convergence of gr
theory, social network theory, and data mining.

Chapter 5

SemiDiscrete Decomposition (SDD)

The singular value decomposition is best thought of as a transformation i
a geometric space. The SemiDiscrete Decomposition (SDD), although it wa
originally developed as a space-efficient version of SVD, is best thought o
as working with the entries of the matrix directly. It searches for regions i
the matrix that have entries of roughly similar magnitude, and treats each o
these as a component. In many situations, this is a useful way to decompos
the matrix.

5.1   Definition

Given a matrix A, the SemiDiscrete Decomposition of A of dimension k is

$$A \approx X_k \, D_k \, Y_k'$$

If A is $n \times m$, then X_k is $n \times k$, D_k is a diagonal $k \times k$ matrix, and Y_k' :
$k \times m$. k does not need to be smaller than m. The entries of X_k and Y_k ca
be only $0, +1$, or -1.

Unlike SVD, where the basic decomposition exactly describes A, th
limitation that X and Y can contain only 0s, 1s or -1s means that the righ
hand side may not exactly describe A, even for quite large k. SDD is force
to generalize the contents of A because of a restriction on the range of valu
rather than a restriction on the size of the decomposition matrices.

Each row of X corresponds to a row of A, and each column of Y' co
responds to a column of A. The diagonal entries of D, like those of S in a
SVD, provide information about the importance of each of the components

Whereas the natural way to understand SVD is as a transformation i geometric space, SDD is best understood as a transformation within the d matrix itself. Suppose that we consider A as an $n \times m$ grid of entries; view each positive entry as a tower at that position in the grid, with a hei proportional to the value of the entry; and we view each negative entry a hole at that position, with a depth that corresponds to the value of the en In other words, suppose that we view A as if it were the downtown area c city.

SDD finds sets of locations on the grid that are rectilinearly align that is which lie on the same rows and columns, and have similar height depth. In other words, SDD looks for blocks, not necessarily adjacent, similar height towers and similar depth holes.

When such a block is found, it defines one component of the decomp tion. The average height/depth of the block is computed, and removed fr all of the towers and holes involved. The process of searching for the n block then continues.

Although there is a heuristic aspect to the algorithm, the goal in e round is to find a block with large squared volume, that is which both cov many entries of the matrix and has large average height/depth.

For example, consider this small example matrix:

$$A = \begin{bmatrix} 2 & 1 & 1 & 8 & 8 \\ 1 & 4 & 4 & 1 & 1 \\ 1 & 4 & 4 & -8 & -8 \end{bmatrix}$$

Figure 5.1 shows a tower/hole view, in which it is clear that there is a bl defined by the last two entries of the first and last rows. Notice that entries forming the block do not have to be adjacent, and how height a depth matter, but sign does not.

The average magnitude of this block is 8. It cannot be expanded include more entries without reducing its squared volume.

The first column of X, the first diagonal entry of D, and the first colu of Y describe this block. The product of this column of X and this row of define an $n \times m$ stencil or footprint describing the locations of this block, w a $+1$ signifying a location where the block is positive, and a -1 signify a location where the block is negative. The diagonal entry of D defines average height of the block. The product of these three pieces is

$$\begin{bmatrix} 1 \\ 0 \\ -1 \end{bmatrix} \times 8 \times \begin{bmatrix} 0 & 0 & 0 & 1 & 1 \end{bmatrix} = \begin{bmatrix} 0 & 0 & 0 & 8 & 8 \\ 0 & 0 & 0 & 0 & 0 \\ 0 & 0 & 0 & -8 & -8 \end{bmatrix}$$

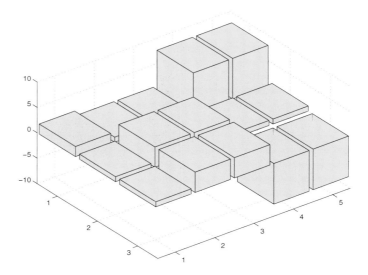

Figure 5.1. *Tower/hole view of the example matrix.*

This block is then subtracted from the original matrix, to produce residual matrix, and the algorithm searches for the next block. The residua matrix is:

$$R = \begin{bmatrix} 2 & 1 & 1 & 0 & 0 \\ 1 & 4 & 4 & 1 & 1 \\ 1 & 4 & 4 & 0 & 0 \end{bmatrix}$$

The next block is the block of 4s, so the next column of X is

$$\begin{bmatrix} 0 \\ 1 \\ 1 \end{bmatrix}$$

the next row of Y' is

$$\begin{bmatrix} 0 & 1 & 1 & 0 & 0 \end{bmatrix}$$

and the next diagonal entry of D has the value 4, the average magnitude c these entries.

Subtracting this block from the residual matrix leaves this new residua

$$R = \begin{bmatrix} 2 & 1 & 1 & 0 & 0 \\ 1 & 0 & 0 & 1 & 1 \\ 1 & 0 & 0 & 0 & 0 \end{bmatrix}$$

Now it becomes hard to see, intuitively, what the best next component is. I fact, the algorithm chooses a stencil covering the first two rows, and all th columns, so the next column of X is

$$\begin{bmatrix} 1 \\ 1 \\ 0 \end{bmatrix}$$

the next row of Y' is

$$\begin{bmatrix} 1 & 1 & 1 & 1 & 1 \end{bmatrix}$$

and the next diagonal entry of D has value 0.7 ($= 2 + 1 + 1 + 0 + 0 + 1 + $
$0 + 1 + 1/10$).

The residual matrix after subtracting this component is:

$$R = \begin{bmatrix} 1.3 & 0.3 & 0.3 & -0.7 & -0.7 \\ 0.3 & -0.7 & -0.7 & 0.3 & 0.3 \\ 1 & 0 & 0 & 0 & 0 \end{bmatrix}$$

and it becomes even more difficult to see intuitively what the next blo
should be. The entries of the residual matrix continue to get smaller in m
nitude as further blocks are removed, so choosing to terminate the decom
sition at some value of k leaves a residual that exactly captures how m
structure is missed by the choice.

In the first two rounds for the example matrix, the average height
each block was the same as the magnitudes of the heights of every entry
the block. Once a block was subtracted from the matrix, the entries in
residual at those locations were zero, so those locations were unlikely to
part of a subsequent block. However, in the later rounds, this was no lon
true, so there are entries in the matrix that are not zero, despite being p
of an already removed block. This is because a region with *average* heigh
removed.

Suppose that a matrix contains a region like this:

$$\begin{matrix} 8 & 8 \\ 8 & 7.5 \end{matrix}$$

and it is selected as a component. The average height of this region is 7
($= (8 + 8 + 8 + 7.5)/4$) so this amount is subtracted from all of the locati
in this region, giving:

$$\begin{matrix} 0.13 & 0.13 \\ 0.13 & -0.37 \end{matrix}$$

Any of these locations may be selected as part of some other block. Inde
exactly this same set of locations may be selected as a block again, but
course with a different height. Hence X and Y can contain identical re
and columns several times, representing 'echoes' of regions that were alrea
removed, but not completely, from the data matrix.

We can see from this example that the general effect of SDD is to find regions of the matrix in which the *volume* is relatively large, in the sense that the region both covers many locations, and contains entries whose magnitude are large. Consider what such a region means in terms of the original dataset. Such a region selects *some* objects and *some* of their values, so in the geometric sense it is a region, in a subset of the dimensions, of homogeneous density. SDD can be considered a form of *bump hunting* [44].

For the matrix we have been using as our running example, the X and Y matrices are:

$$X = \begin{bmatrix} -1 & 0 & -1 & 1 & -1 & 0 & 0 & -1 \\ -1 & 0 & 0 & 1 & -1 & 0 & 1 & 0 \\ -1 & 1 & 1 & 0 & -1 & -1 & 0 & -1 \\ 1 & 1 & 1 & 0 & 0 & -1 & 1 & 1 \\ 1 & -1 & -1 & 1 & 0 & -1 & 1 & 1 \\ -1 & 0 & 0 & -1 & 1 & 0 & 1 & -1 \\ 0 & 0 & 0 & -1 & -1 & 0 & 0 & -1 \\ 0 & -1 & 0 & -1 & 1 & 0 & 1 & 0 \\ 0 & 0 & 0 & 0 & 1 & -1 & -1 & 0 \\ 0 & -1 & 1 & 1 & 0 & 1 & 0 & 0 \\ 0 & 1 & -1 & 0 & -1 & 0 & -1 & 1 \end{bmatrix}$$

$$Y = \begin{bmatrix} 1 & 0 & 0 & 0 & 0 & -1 & 1 & 0 \\ 0 & 1 & 1 & 0 & 0 & -1 & -1 & 1 \\ 1 & 0 & 0 & 1 & 0 & 0 & 1 & 1 \\ 0 & 1 & 1 & 1 & 0 & 0 & 1 & -1 \\ 1 & 1 & -1 & 1 & 0 & 0 & -1 & 0 \\ -1 & 0 & 1 & 1 & 1 & 0 & -1 & -1 \\ 0 & -1 & 0 & 1 & -1 & 1 & 1 & 0 \\ -1 & 1 & -1 & 0 & 0 & 1 & 1 & 1 \end{bmatrix}$$

and the values of D are

$$D = \begin{bmatrix} 0.87 \\ 0.61 \\ 0.68 \\ 0.56 \\ 0.51 \\ 0.46 \\ 0.22 \\ 0.30 \end{bmatrix}$$

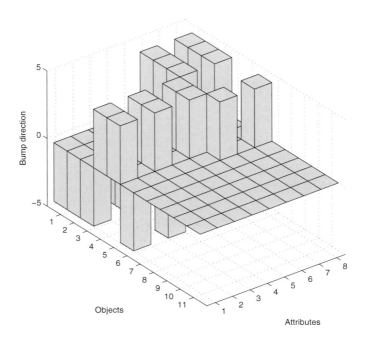

Figure 5.2. *Bumps at level 1 for the example matrix.*

The bump at the second level is:

$$
=
\begin{bmatrix}
0 & 0 & 0 & 0 & 0 & 0 & 0 & 0 \\
0 & 0 & 0 & 0 & 0 & 0 & 0 & 0 \\
0 & 1 & 0 & 1 & 1 & 0 & -1 & 1 \\
0 & 1 & 0 & 1 & 1 & 0 & -1 & 1 \\
0 & -1 & 0 & -1 & -1 & 0 & 1 & -1 \\
0 & 0 & 0 & 0 & 0 & 0 & 0 & 0 \\
0 & 0 & 0 & 0 & 0 & 0 & 0 & 0 \\
0 & -1 & 0 & -1 & -1 & 0 & 1 & -1 \\
0 & 0 & 0 & 0 & 0 & 0 & 0 & 0 \\
0 & -1 & 0 & -1 & -1 & 0 & 1 & -1 \\
0 & 1 & 0 & 1 & 1 & 0 & -1 & 1
\end{bmatrix}
$$

The bumps at the first few levels for the example matrix are shown
Figures 5.2, 5.3, and 5.4.

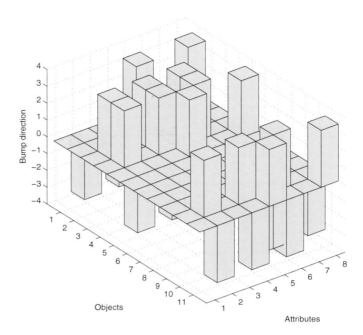

Figure 5.3. *Bumps at level 2 for the example matrix.*

Normalization

SDD is a numerical technique so it is necessary to give some thought t
the relative magnitudes associated with each attribute (column). Values c
comparable significance should have comparable magnitudes. For an arbitrar
dataset this may be problematic, at least without some understanding of th
problem domain. It might be plausible, for example, to divide the entrie
in each column by the maximum entry to bring all attributes into a simila
range. There are some situations where the attributes are all of the sam
kind, for example word frequencies in documents, and SDD can be applie
more straightforwardly then.

SDD does not require that attribute values be centered around the origir
so if the data is, for example, naturally non-negative then SDD can be applie
directly. However, for some forms of data, it may be the extremal values c
attributes that are most significant. In this case, it may be appropriate t
zero center each attribute by dividing each entry by the attribute mean.

For example, suppose that A is a dataset from a collaborative filterin
setting, so that the rows of A correspond to individuals, the columns of .
correspond to objects being rated, and each entry is a rating of an object b

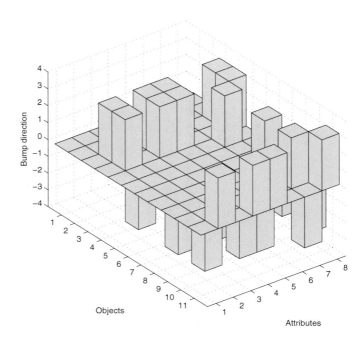

Figure 5.4. *Bumps at level 3 for the example matrix.*

an individual, say on a scale from 1 to 10. Since each column represents
same kind of attribute, and they all have the same range of magnitudes, th
is no need to scale the columns. Applying SDD to this matrix selects grou
of objects that receive high ratings from significant numbers of individuals:
other words it selects good (highly rated) objects.

Zero centering the columns, and then applying SDD has a different eff
– it selects groups of objects that receive strongly positive or negative ratin
from significant numbers of individuals. In other words, it selects obje
about which opinions are strong. If the objects are movies, then the fi
analysis may select movies that will make money, while the second analy
may select movies that will receive critical acclaim.

An SDD is not scale-independent because the order in which bumps
selected depends on heights as well as areas. Altering the relative magnitu
of these two factors can change the order of selection. For example, if t
magnitudes of the entries of a matrix are scaled up by squaring each one a
preserving its sign, then the effect is to select small regions of unusually la
magnitude first. If the entries are replaced by their signed square roots, t
effect is to select large regions of smaller magnitude first.

Reordering bump selection

The heuristic embedded in the basic SDD step creates two problems, one of them solvable, the other not. The algorithm is quite sensitive to the initial choice of y_i. This means that it does not always find the largest possible bump to remove from the matrix at each step. Hence later steps can find large bump that was missed on previous steps. As a result, the values on the diagonal of D are not always decreasing.

We apply the following modification to the algorithm. After the X, Y and D matrices have been computed, we

1. Form the product of the d_is with the number of non-zero entries in the corresponding columns of Y, and

2. Sort the columns of X, elements of D and rows of Y into decreasing order of the products from the first step.

This has the effect of reordering the bumps so that those with the largest volume appear first. In other words, the strongest components appear in the earliest columns of X.

The second problem occurs because the height of a bump removed depends on the current contents of the matrix, which depends on the order in which previous bumps were removed. Reordering at the end cannot reproduce exactly the effect of having chosen a different removal order during the algorithm's execution. The problem occurs because the height of a bump determined by the *average* height of the locations that will be removed. In some fundamental sense, the way in which the d_i are computed prevents the original matrix from being partitioned as cleanly as it might otherwise be. There seems to be no simple solution to this problem.

For our running example, the X and Y matrices become:

$$X = \begin{bmatrix} -1 & -1 & 0 & 1 & 0 & 0 & -1 & -1 \\ -1 & 0 & 0 & 1 & 0 & 1 & 0 & -1 \\ -1 & 1 & 1 & 0 & -1 & 0 & -1 & -1 \\ 1 & 1 & 1 & 0 & -1 & 1 & 1 & 0 \\ 1 & -1 & -1 & 1 & -1 & 1 & 1 & 0 \\ -1 & 0 & 0 & -1 & 0 & 1 & -1 & 1 \\ 0 & 0 & 0 & -1 & 0 & 0 & -1 & -1 \\ 0 & 0 & -1 & -1 & 0 & 1 & 0 & 1 \\ 0 & 0 & 0 & 0 & -1 & -1 & 0 & 1 \\ 0 & 1 & -1 & 1 & 1 & 0 & 0 & 0 \\ 0 & -1 & 1 & 0 & 0 & -1 & 1 & -1 \end{bmatrix}$$

$$
Y = \begin{bmatrix}
1 & 0 & 0 & 0 & -1 & 1 & 0 & 0 \\
0 & 1 & 1 & 0 & -1 & -1 & 1 & 0 \\
1 & 0 & 0 & 1 & 0 & 1 & 1 & 0 \\
0 & 1 & 1 & 1 & 0 & 1 & -1 & 0 \\
1 & -1 & 1 & 1 & 0 & -1 & 0 & 0 \\
-1 & 1 & 0 & 1 & 0 & -1 & -1 & 1 \\
0 & 0 & -1 & 1 & 1 & 1 & 0 & -1 \\
-1 & -1 & 1 & 0 & 1 & 1 & 1 & 0
\end{bmatrix}
$$

when the reordering version of the algorithm is used, and the values of D now:

$$
D = \begin{bmatrix}
4.33 \\
3.38 \\
3.03 \\
2.81 \\
1.85 \\
1.77 \\
1.52 \\
1.02
\end{bmatrix}
$$

The bump at the second level is now:

$$
A_2 = \begin{bmatrix}
0 & -1 & 0 & -1 & 1 & -1 & 0 & 1 \\
0 & 0 & 0 & 0 & 0 & 0 & 0 & 0 \\
0 & 1 & 0 & 1 & -1 & 1 & 0 & -1 \\
0 & 1 & 0 & 1 & -1 & 1 & 0 & -1 \\
0 & -1 & 0 & -1 & 1 & -1 & 0 & 1 \\
0 & 0 & 0 & 0 & 0 & 0 & 0 & 0 \\
0 & 0 & 0 & 0 & 0 & 0 & 0 & 0 \\
0 & 0 & 0 & 0 & 0 & 0 & 0 & 0 \\
0 & 0 & 0 & 0 & 0 & 0 & 0 & 0 \\
0 & 1 & 0 & 1 & -1 & 1 & 0 & -1 \\
0 & -1 & 0 & -1 & 1 & -1 & 0 & 1
\end{bmatrix}
$$

representing a different set of locations to those selected by the original al rithm.

5.2   Interpreting an SDD

From the discussion above, it is clear that the component interpretation the most natural one for an SDD. However, it is worth considering the ot interpretations as well, at least for some datasets.

Unlike SVD, the pieces of an SDD do not directly represent either obje or attributes, but combinations of them – in other words, the decomposit

inherently reveals structure in *some* attributes of *some* objects. As discusse
in Chapter 1, this is, in many ways, an advantage because many real datase
do not have structure related to *all* of the attributes, or *all* of the object
However, this does make interpretation more difficult.

5.2.1 Factor interpretation

In the factor interpretation, the rows of Y' are interpreted as factors tha
are mixed by the rows of X and diagonal entries of D. There are sever
limitations to this interpretation. First, the mixing is an all or nothing matte
since the entries of X are either 0 or $-1, +1$. Second, the rows of Y' are n
independent; as we have already observed, the same row can occur sever
times as a result of the echo effect. Third, for most problems, limiting th
entries of Y' to $0, -1$, or $+1$ forces too simple a representation for the 'rea
factors.

For example, suppose that a matrix contains a block of locations wit
values 40, 40, 40, and 20, and that all of the other values are much smalle
The bump that is removed because of this block will have $d_i = 35$ (the averag
magnitude of these locations). The next bump may well be the correctic
needed for the location with value 20, this time a bump of height 15, but i
the negative direction ($20 - 35 = -15$). In other words, the second bum
is a correction for part of the first, and so these two bumps are not, in an
reasonable sense, independent factors.

There are settings where simple factors are appropriate, notably imag
processing, which is in fact the domain where SDD was developed. Here th
factors might represent pixels, pieces, or even objects in an image, and simp
mixtures are appropriate.

5.2.2 Geometric interpretation

SDD has a form of geometric interpretation. The rows of Y can be regarde
as defining 'generalized quadrants'; the values of X then specify whether
given object is placed in a given 'quadrant' or not.

Each 'bump' in an SDD has a natural geometric interpretation in th
geometric space corresponding to the original matrix, A. Such a bump is
region of unusual, homogeneous values in a subset of the original dimension
In other words, each bump delineates a submanifold in the original space (i
fact a generalized cube since the coordinate values in the dimensions involve
are of roughly the same magnitude). Discovering such submanifolds can t
difficult, so one useful property of SDD is that it finds groups of attribut
that play consistent roles in representing the data.

The bumps are in fact a kind of clustering, but one which clusters objects and, at the same time, clusters the attributes. Such clusterings known as *biclusterings*.

5.2.3 Component interpretation

The component interpretation is the natural one for SDD. Each entry in array A can be expressed as the sum of A_is, where each A_i is the produc a column of X, an entry in D, and a row of Y (that is, a bump). As w SVD, it is possible to use only part of this sum, and terms associated w large values of D are more significant.

Unlike components in an SVD, which typically represent global proce and so affect every entry of the data matrix, SDD components are lo each one affecting only some entries of the data matrix. SDD is there an appropriate decomposition when the processes that were combined in data matrix have this same local character. For example, noise tends to a global phenomenon, so we would expect it to alter all of the entries dataset. SDD cannot be expected to be effective at removing noise of kind.

5.2.4 Graph interpretation

The graph interpretation for SDD does not seem to produce any new insig because the non-zero entries correspond only to the existence of edges. the resulting tripartite graph, the number of edges incident at each 'mid vertex gives the dimensions of the stencil corresponding to it. The num of edges leaving each vertex corresponding to an object or attribute descri how many bumps it participates in and, when the edges are weighted by diagonal entries of D, how significant each is. This information is rea available from the decomposition, however.

5.3 Applying an SDD

5.3.1 Truncation

Because the components of an SDD are selected based on volume, and cc or explain a limited number of values in the data matrix, there is no natu way to remove noise by eliminating some components.

It might be reasonable to select or discard certain components ba on the shape and location of their footprint, and their height. However, would almost certainly require some domain knowledge – the decomposit itself does not provide much helpful information.

5.3.2   Similarity and clustering

We have already seen that SDD components correspond to clusters in the geometric space of the data matrix. Such clusters are actually biclusters since each one is based only on a subset of the attributes. For many datasets this is more appropriate than a clustering based on all of the attributes.

This clustering is like a partitional clustering, that is with all clusters treated as being on the same level, but it does not necessarily partition the objects. Some objects may not appear in any cluster, while others might be members of several: either distinct clusters, or an original cluster and its echoes.

Hierarchical clustering

However, the SDD actually imposes an ordering on the bumps, which can be made into an ordering on the clusters, producing a hierarchical clustering.

The first column of X can be thought of as dividing the data objects into three kinds: those for which the entry in X is +1, those for which it is 0, and those for which it is −1. Now for each of these three groups, we can consider the entries of the second column of X. These subdivide each group into three subgroups. This process can be continued for subsequent columns of X.

The result is a hierarchical clustering of the objects of A. However, it is an unusual clustering. First, the partitions at different levels are technically independent, since they describe partitions in different components, and there is no necessary relationship between different components. However, levels do have an importance ordering, so the hierarchical clustering does say something about the role of each object in different components. This is a weak criterion, so it is always possible to interchange levels and get a different hierarchical clustering. Second, unlike a standard hierarchical clustering, the result is a ternary tree rather than the more usual binary tree of a dendrogram. Third, the −1 and +1 branches are 'equal and opposite' rather than 'different'. Fourth, the tree is constructed in a top-down way, rather than the bottom-up construction of a typical dendrogram.

The resulting hierarchical clustering has k levels, and each node at level l is characterized by a string of −1s, 0s, and +1s of length l. The parent label of any node is obtained by removing the final symbol from its own label. Not every node of the tree will necessarily be populated, and it may be convenient to truncate a branch of the tree at the point when the leaf node contains only a single object.

The depth of a node in the tree is not as significant as in a dendrogram because the 0-labelled branch at each level implies no additional information

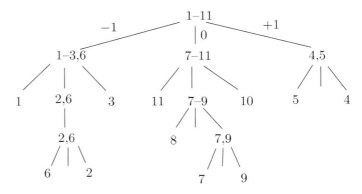

Figure 5.5. *Hierarchical clustering for objects of the example matrix.*

about the objects concerned. For example, if the first two components single outlying objects, then almost all objects will lie below the node label '00', but this does not make them somehow less important than if the outly objects had not been present. In other words, SDD provides information ab similarity, but provides only weak information about dissimilarity.

The hierarchical clustering for the example matrix is shown in Figure 5.5.

The same procedure can be followed, using Y, to build a hierarch clustering of the attributes of A.

Measuring similarity in hierarchical clusterings

In order to determine similarity between objects, we need to impose a dista measure on the hierarchical clustering generated from X or Y. We can use the labels on the edges ($+1$ and -1) directly because branches with th labels at the same level represent clusters that are 'similar but opposite', a so should be counted as close together.

A similarity metric that seems to be useful is to count a distance of

- 0 for traversal of a zero-labelled branch of the tree (because the existe of clusters at other levels between two clusters of interest shouldn't m them seem more different);

- $+1$ for traversal of a $+1$-labelled or -1-labelled branch (because t represents moving to or from a bump); except ...

- $+1$ for traversal between the $+1$-labelled and -1-labelled branches *the same level* (because they are equal but opposite bumps).

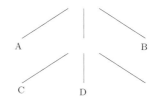

Figure 5.6. *Examples of distances (similarities) in a hierarchical clustering.*

In Figure 5.6, the distance between A and B is 1 because these nodes of the tree represent towers and holes from the same bump. The distance between C and B is 2, one step to get on to the main 'trunk' of the tree, and another step to get off again. The distance between D and B is 1 because objects in cluster D are undifferentiated (have not been members of any bump). These objects are dissimilar to those of cluster B, but we cannot tell from the tree how much.

This distance measure can be used to decide on the relationships among clusters of objects and attributes, in the same way that proximity in low dimensional space was used in SVD.

Selecting outliers

There are a number of ways of using SDD to decide which objects (or attributes) are outliers. For example, nodes in the hierarchical clustering that contain one, or just a few, objects might be considered to describe outliers. However, if the hierarchical clustering is continued far enough then every object is in a node by itself (see Figure 5.5), so only nodes with small numbers of objects *near the root of the tree* are really outliers. Such nodes describe rather obvious outliers since the objects they describe must have a number of attributes with unusual values to have been selected near the root.

Another way to characterize outliers that are somehow extremal or most dissimilar to others is that they participate in many bumps. Such objects have attribute values that overlap with other common attribute values, but not enough to enable them to be well-explained by a single, or even a few, bumps. To say it another way, an object that correlates well with many others will tend to be placed in a bump or a small number of bumps of reasonable size with them, and so its attribute values will come from a few components.

We define the following 'bumpiness' score for an object i to capture this intuition

$$\text{bumpiness score}(i) \ = \ \sum D(i) \times abs(X(i))$$

taking the ith row of X and the ith diagonal element of D. Objects wit
large bumpiness score are considered as outliers. (Replacing X by Y give
bumpiness score for attributes.)

Applying SDD to correlation matrices

SDD can be especially effective when applied to correlation matrices, AA'
$A'A$. Whenever a set of objects have similar attributes, the dot product
their rows will be of a similar magnitude and sign, and these dot produ
form the entries of the correlation matrix, AA'. Hence a block of sim
magnitude in a correlation matrix corresponds exactly to a set of sim
objects in the original data matrix. Because SDD also selects blocks ba
on regions with similar negative magnitudes, it also finds objects that
negatively correlated. Of course, exactly the same thing applies to correlati
among attributes, which form blocks in the matrix $A'A$.

5.4   Algorithm issues

As the informal description of the algorithm suggested, an SDD is built
iteratively, one component at a time. Let x_i be the ith column of X, d_i
ith diagonal element of D, and y_i the ith row of Y'. The standard algorit
for computing the SDD generates a new column, diagonal element, and
on each step. Let A_0 be the $n \times m$ matrix of zeroes. The algorithm is,
each step i

1. Subtract the current approximation, A_{i-1}, from A to get a resid
   matrix R_i.

2. Find a triple (x_i, d_i, y_i) that minimizes

$$\| R_i - d_i x_i y_i \|^2 \qquad (*)$$

   where x_i is $n \times 1$ and y_i is $1 \times m$. The standard algorithm uses
   following heuristic:

   (a) Choose an initial y_i.
   (b) Solve $(*)$ for x_i and d_i using this y_i.
   (c) Solve $(*)$ for y_i and d_i using the x_i from the previous step.
   (d) Repeat until some convergence criterion is satisfied.

3. Repeat until $i = k$.

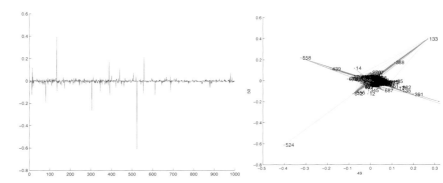

Figure 1. *Eigenvector and graph plots for column 50 of the U matrix.*

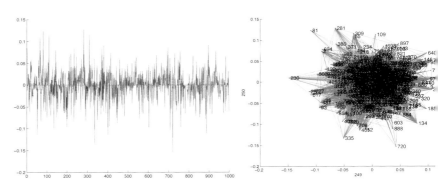

Figure 2. *Eigenvector and graph plots for column 250 of the U matrix.*

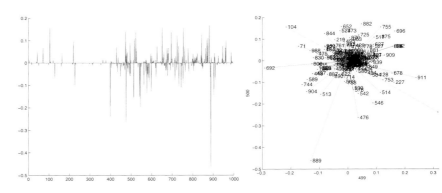

Figure 3. *Eigenvector and graph plots for column 500 of the U matrix.*

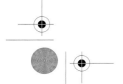

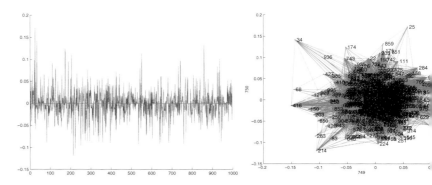

Figure 4. *Eigenvector and graph plots for colum 750 of the U matrix.*

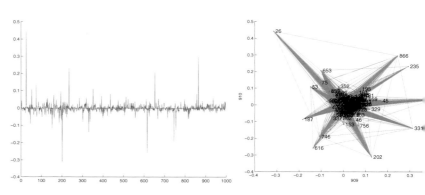

Figure 5. *Eigenvector and graph plots for column 910 of the U matrix.*

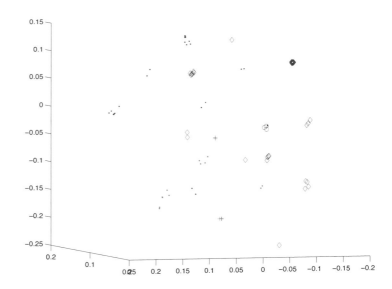

Figure 6. *Plot of sparse clusters, position from the SVD, shape (most significant) and co* the SDD.

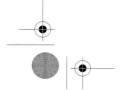

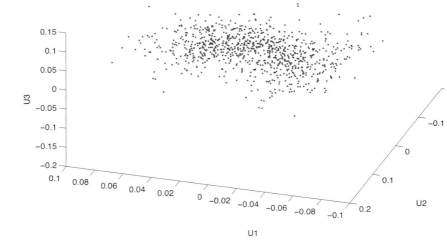

Figure 7. *Plot of an SVD of galaxy data.*

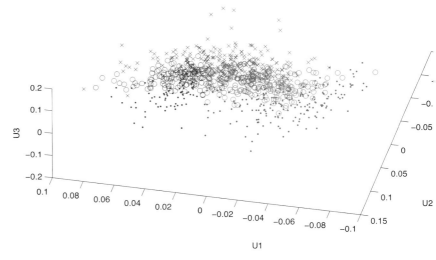

Figure 8. *Plot of the SVD of galaxy data, overlaid with the SDD classification.*

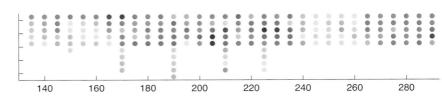

Figure 9. *pH by sample; darker color means lower pH, greater acidity.*

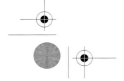

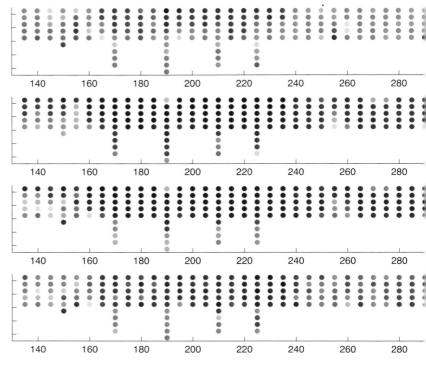

Figure 10. *Calcium concentration by digestion (AA5, AA7, AQR, GDX).*

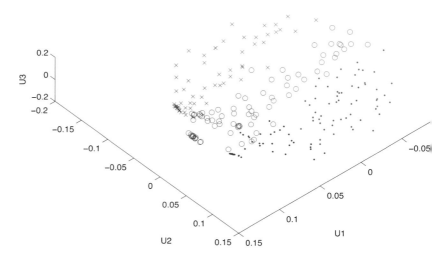

Figure 11. *Plot with position from the SVD, and color and shape labelling from the SI*

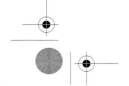

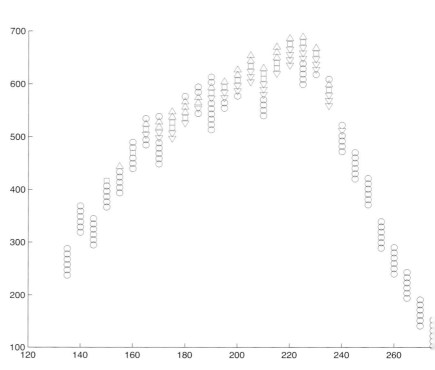

Figure 12. *Sample locations labelled using the top two levels of the SDD classification: up triangle = 1, −1, square = 0, −1, downward triangle = −1, −1.*

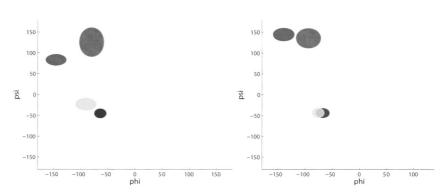

Figure 13. *(a) Conformations of the ASP-VAL bond; (b) Conformations of the VAL-ALA*

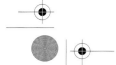

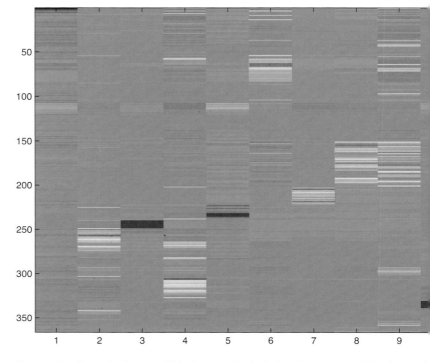

Figure 14. *C matrix from an ICA of a matrix of relationships among al Qaeda memb*

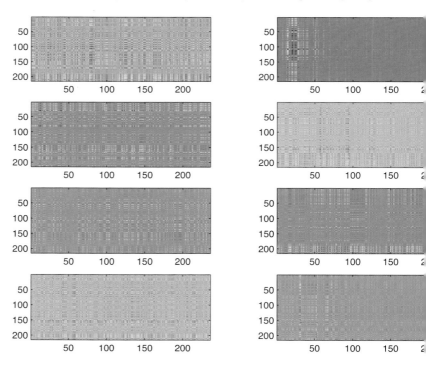

Figure 15. *Outer product plots for the SVD.*

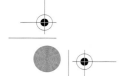

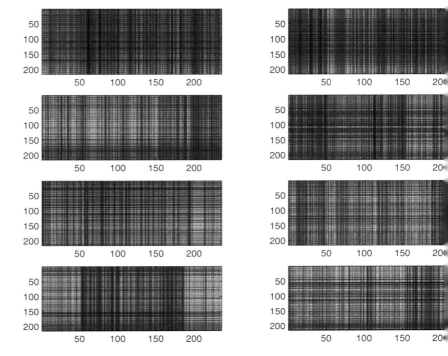

Figure 16. *Outer product plots for Seung and Lee's NNMF.*

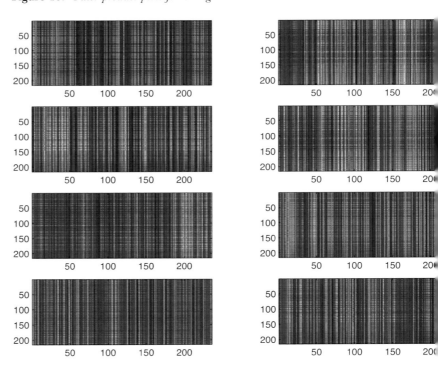

Figure 17. *Outer product plots for Gradient Descent Conjugate Least Squares NNMF.*

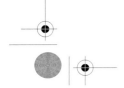

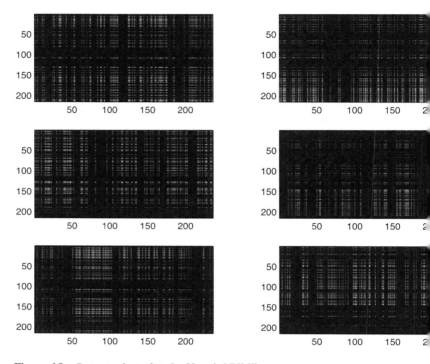

Figure 18. *Outer product plots for Hoyer's NNMF.*

Complexity

Matlab and C code for this algorithm is available from `www.cs.umd.edu/user`
`oleary/SDDPACK/`. Because the algorithm is a heuristic, some datasets wi
require changes to the default parameters, which control the initial choice c
y_i and the convergence criteria.

The complexity of the SDD heuristic algorithm is $\mathcal{O}(k^2(n+m)+n\log n$
$m\log m)$, comparable to SVD if we choose $k = m$.

Exact algorithm versus heuristic

We have described a heuristic for computing the SDD which greedily selects
candidate column for X and Y at each iteration. This algorithm is generall
fairly stable, although the way in which the initial y_j vector is determine
may have to be adjusted for some datasets. An exact algorithm could als
be used, but its complexity is $\mathcal{O}((n+m)^3)$ so this is useful only for sma
datasets.

The only reason for choosing a measure for bump volume that include
a quadratic term is that it creates a minimization problem with an easy algc
rithm. There are several plausible extensions: using a criterion that represent
the usual volume of a bump; and using the magnitude, rather than the al
solute value of the magnitude, so that positive and negative entries do nc
behave as if they are related.

Removing parts of bumps

One of the weaknesses of SDD is that a peak in the dataset with (say) tw
distinct values tends to be selected as a single bump whose height is th
average of the two data values. Interesting patterns can sometimes be see
by selecting bumps as in the basic algorithm, but removing only a part (sa
half) of their height. This may result in the same bump being removed in tw
stages; but it can also allow the remaining part of a bump to be seen as par
of some larger bump.

5.5   Extensions

5.5.1   Binary nonorthogonal matrix decomposition

Grama [77] has proposed a kind of generalization to SDD that decompose
binary sparse datasets using outer products that contain only 0s and 1s (s
without −1s as in SDD). The overall structure of the algorithm is similar t

SDD – at each stage, a column and row vector are found whose outer prod
is close to the current matrix in the sense of Hamming distance (i.e., numbe
non-zeroes). However, rather in the style of PDDP, the dataset is partitio
at each stage based on whether the column vector contains 0s or 1s, and
process is repeated for each submatrix. Hence the original dataset can
expressed as a sum of components, but this sum is a tree reduction rat
than a list reduction as it is for SVD and SDD.

Notes

The SemiDiscrete Decomposition was initially developed by Peleg and O'Le
[90], and developed by O'Leary and Kolda [70, 72, 73, 119]. Its characteri
tion as a bump-hunting technique can be found in [85].

Chapter 6

Using SVD and SDD together

Although SVD and SDD work in different ways, if a dataset contains a genuine clustering, it should be visible to both algorithms. SVD and SDD are quite complementary. SVD is able to make the most important structure visible in the early dimensions, but it is hard to exploit this directly because there are multiple ways to construct and label clusters from it. SDD, on the other hand, tends to produce more, smaller clusters than SVD (because they are really biclusters) but provides an automatic labelling of objects with the cluster they belong to, using some subset of the columns of X.

Using both decompositions together can often provide insights that neither could produce on its own. This is especially true when visualization is used. SDD produces and labels clusters but provides no natural way to visualize them. SVD provides a way to visualize clusters, but no simple way to label them (and, in particular, to delineate the cluster boundaries cleanly).

SVD and SDD tend to agree about the clustering of a dataset when that clustering consists of many, small, well-separated clusters. This is typical, for example, of the clustering of document-word matrices, which explains why SDD has been used effectively as a supplement to LSI. In this case, the benefit of adding SDD analysis to the SVD analysis is that we get a hierarchical clustering of the data. When the clustering consists of a few, large clusters, there is typically much more disagreement between the two decompositions. The advantage of using SDD here is that it provides cluster boundaries within what the SVD considers single clusters, and it can help to decide which cluster outlying or remote objects might best be allocated to.

An example is shown in Figure 6.1, using a dataset typical of text retrieval applications – many zeroes, and the remaining values small positive integers. Here the position of points (in 3 dimensions) is determined by the

141

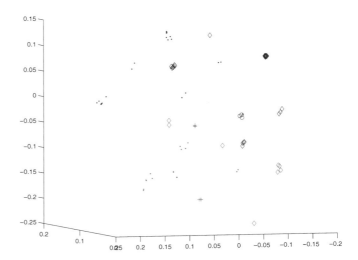

Figure 6.1. *Plot of sparse clusters, position from the SVD, sh*
(most significant) and color from the SDD. (See also Color Figure 6 in
insert following page 138.)

SVD but the shape and color used to render them is determined by the SI
It is clear that each cluster (in the SVD sense) is homogeneously labelled
the SDD. Hence the two techniques agree.

Figure 6.2 shows the SDD hierarchical clustering label for each object in
running example dataset, superimposed on a plotted position from the S
Figure 6.3 shows the same kind of plot for the attributes. There is gen
agreement about classification, but there are some differences – together S
and the SDD provide different views of the same data.

6.1   SVD then SDD

In the previous section, we considered SVD and SDD analysis in parallel. N
we consider how the two decompositions can be used in series. In particu.
an SVD can be used to clean the original dataset, making it easier for an SI
to detect bumps that may be present in the data.

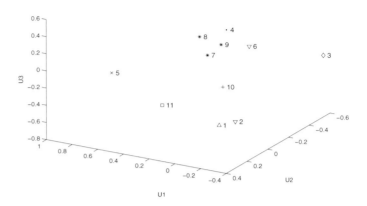

Figure 6.2. *Plot of objects, with position from the SVD and labellir from the SDD; symbol shapes describe the branching at the first two levels the decision tree like this:* $\cdot$ $1, 1$, o $1, 0$, $\times$ $1, -1$, $+$ $0, 1$, $*$ $0, 0$, $\diamond$ $-1, 1$, $-1, 0$, $\triangle$ $-1, -1$.

6.1.1 Applying SDD to A_k

The simplest way to combine SVD and SDD sequentially is to compute th SVD of the data matrix, truncate it at some suitable k, remultiply to produc a matrix A_k of the same size as A, and then decompose A_k using an SDD.

The logic of this combination is that the SVD is denoising the dat matrix so that the SDD can better see the structure within it. The labels fro the early columns of X can be used as labels for the clusters that the SD finds. These labels can be used to overlay the positions of points correspondin to the objects from the early columns of U_k. Although this seems like straightforward application of SVD, the combination is remarkably effectiv – the effect of the SVD is to tighten up the boundaries of clusters present i the data, and the effect of the SDD is to identify these clusters accurately.

6.1.2 Applying SDD to the truncated correlation matrices

For some datasets, a further improvement can be made by adding the follow ing refinement: first, perform an SVD on the data matrix, truncate as befor at some k, and remultiply to create a matrix, A_k. Now form the correla tion matrix $A_k A_k'$ and apply the SDD to this new correlation matrix. Plc points corresponding to the coordinates in U_k, and label them according t

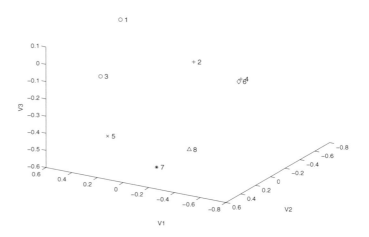

Figure 6.3. *Plot of attributes, with position from the SVD and belling from the SDD.*

the hierarchical clustering applied to the truncated correlation matrix. same process can, of course, be applied to the attributes using the attrib correlation matrix, $A_k' A_k$.

The logic of this combination is that the SVD is denoising the data trix to make its structure clearer, but then the SDD is used to find the co lation structure within the denoised data, rather than finding the magnit structure. We call this combination the JSS (Joint SDD-SVD) methodolc

Since the diagonal of a correlation matrix tends to contain large val that are not directly relevant to understanding the correlation structure a matrix, and because a diagonally oriented set of similar values is hard SDD to represent, we set the diagonal values of correlation matrices to z There is probably a more intelligent way to address this issue.

6.2   Applications of SVD and SDD together

6.2.1   Classifying galaxies

Figure 6.4 is a plot from the U matrix of an SVD of a dataset containing d about 863 galaxies, with attributes that are corrected intensities at a set four frequencies. The first three columns provide coordinates in three dim sions for each galaxy. Hubble proposed a classification for galaxies into th types: ellipticals, spirals (now subdivided into spirals and barred spirals),

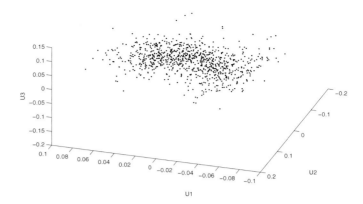

Figure 6.4. *Plot of an SVD of galaxy data. (See also Color Figure 7 in the insert following page 138.)*

irregulars. There is little indication of any clustering in the Figure that might agree with this classification.

Figure 6.5 shows the same data with the position from the first three columns of U as before, but the color and shape of the symbols based on the SDD clustering. Color is based on the first column of X (red = +1, green = 0, blue = −1), and the shape is based on the next column of X, with · = +1, o = 0, and × = −1. Although this dataset is almost certainly not rich enough to classify galaxies using the Hubble classification, the Figure shows how the SDD provides extra information. This is of two kinds: first, it defines a finer clustering than that of SVD, which basically sees a single large cluster. Second, it provides a way to decide on the role of the stray points that are far from the main SVD cluster. For this dataset, it seems clear that none of the objects are outliers, although this is not obvious from the SVD plot alone.

6.2.2 Mineral exploration

Traditional mineral exploration involves boring deep holes to look for regions under the surface containing desirable minerals, such as copper, zinc, gold, or silver. A much simpler approach that is being developed is to take samples at or near the surface, but over a much wider area. These near-surface samples can be processed to estimate the presence and concentration of a large number of elements. It may be possible to detect a buried region that contains interesting minerals from its effects on the near-surface concentrations

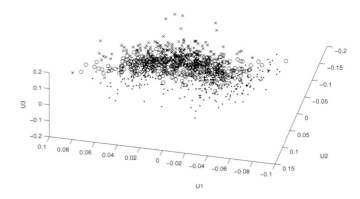

Figure 6.5. *Plot of the SVD of galaxy data, overlaid with the S* *classification. (See also Color Figure 8 in the insert following page 138.)*

and chemistry of these other elements. The near-surface samples are p cessed using several leaches ('digestions') that extract elements from differ compounds, depending on how tightly they are bound in each one. He the concentration of a particular element measured using different digesti gives a contextualized estimate of its presence in the sample.

It is not completely understood how a deeply buried region of m eralization affects the surface geochemistry and there are, of course, ma confounding factors such as agriculture, moisture, parent minerals, and ot geological processes. Smee [106, 107] has suggested that H^+ migration is most important signature-creating process, at least in moist environmer so that a pH low (acidity) might be expected over mineralization. Oth (Hamilton [49]) have suggested that the presence of mineralization crea a reduced column, that in turn produces a pH low near or above the wa table due to redox reactions involving oxygen diffusing downward from surface. Such a column might create the most distinctive signature in annulus *around* the mineralization.

We explore these ideas using a dataset collected near Timmins, in nor western Ontario, along a line that crosses zinc-copper mineralization, cove by 25–40m of clay. Samples were taken along the line at 5m intervals a at depths that varied from 10–110cm. Figure 6.6 shows the position of samples with 20× vertical exaggeration. The mineralization lies beneath region from distance markers 185m to 205m along the sample line. Five dig tions were applied to each sample (ENZ, AA5, AA7, AQR, and GDX) and concentrations of many relevant elements measured. This produced a mat dataset with 215 rows (1 per sample), and 236 columns (1 per digesti element pair).

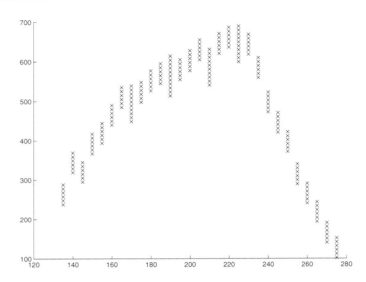

Figure 6.6. *Position of samples along the sample line (some vertice exaggeration).*

Conventional data-mining techniques applied to this dataset did no perform well. Both EM and k-means produced clusters that did not seem t have any physical significance. A decision tree, trained to predict whether sample was over the mineralization or not, predicted that all samples wer not over mineralization. The problem with these techniques is that they trea each sample as if it were completely independent. However, we know that th property of being a sample over mineralization is strongly spatially correlatec and much of the geochemistry probably is too. Adding physical coordinate improves the performance of these techniques a little but, of course, prevent them from being generalized for use on other data.

Color Figure 9, in the insert following page 138, shows the pH by po sition, with lower pH indicated by darker colors. This suggests that pH indeed a good indicator of mineralization, at least in the Canadian Shiel This region of low pH should lead to a calcium depletion near the surface extending down below the surface until oxygen levels become too depleted.

Color Figure 10, in the insert following page 138, shows that this indeed the case, with the depletion zone extending beyond the edges of th mineralization as expected, given a reduced column. So both pH and calciur concentration are good predictors of mineralization in this data, but woul not necessarily generalize to other data, especially when the overburden is n clay.

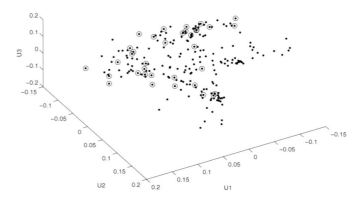

Figure 6.7. *SVD plot in 3 dimensions, with samples over mine
ization circled.*

Figure 6.7 shows a plot of the first three dimensions from an SVD
the dataset, with the samples that lie over mineralization circled. There is
particularly visible clustering structure for these samples – they seem wid
scattered.

The same plot with the SDD hierarchical classification information ov
laid is more revealing (Figure 6.8). The top-level classification is related
depth of sample, with blue indicating the deepest samples while red indica
samples near the surface. The second level of classification is completely
thogonal to the first, indicated by the symbols ×, o, and ·. Many of
samples over mineralization are those denoted by ×.

With the hints from the single-element concentrations that there i
signature associated with an annular region around the mineralization, a
that depth is a significant fact, we consider those samples in the range 1
240m and whose depths are 60cm or less. These points are circled in the p
in Figure 6.9.

This plot still does not seem to contain a cluster whose properties co
be used to predict the presence of mineralization. However, if we consi
Figures 6.8 and 6.9 together, they suggest that there is a well-defined clus
of samples between 150–240m but with a more-constrained range of dept
If we plot the SVD, circling those points that are classified as −1 using SD
we obtain the plot in Figure 6.10, which shows a coherent region with resp
to both classifications.

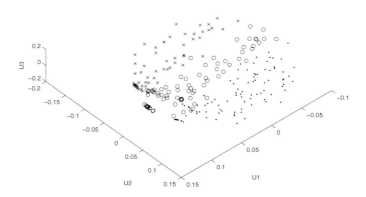

Figure 6.8. *Plot with position from the SVD, and color and shape labelling from the SDD. (See also Color Figure 11 in the insert following page 138.)*

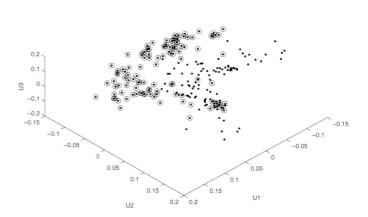

Figure 6.9. *SVD plot with samples between 150–240m and depth less than 60cm.*

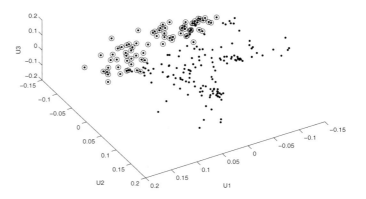

Figure 6.10.   *SVD plot of 3 dimensions, overlaid with the S classification −1 at the second level.*

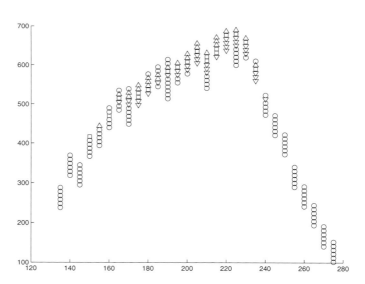

Figure 6.11.   *Sample locations labelled using the top two levels the SDD classification: upward triangle = 1, −1, square = 0, −1, downw triangle = −1, −1. (See also Color Figure 12 in the insert following p 138.)*

Furthermore, this set of samples form a coherent region along the sample line, as shown in Figure 6.11. The effect of depth can be clearly seen, with the first dimension of the SDD classification signalling depth. Note that the deeper samples are not predictive, even when they lie over mineralization. The predictive signature, then, is that seen in a region that lies over, but extends beyond, the actual area of mineralization, and at a depth where both oxygen and water are available.

The dataset does not contain information about sample depth directly. The predictive power of depth has been inferred by the decompositions from its effect on element concentrations. The two decompositions developed here can play the role of predictors, since new samples from elsewhere can be mapped to coordinates and classifications, as long as they are processed using the same digestions, so that the attributes remain the same.

The success of these matrix decompositions in detecting a signature for the presence of mineralization is largely due to their ability to extract correlation information from the various partial element concentrations. Very little information can be extracted from the concentrations of one, or even few, elements. This analysis also provides support for the model of a reduced column above mineralization, with effects extending beyond the area above the mineralization itself. Further details about applying matrix decomposition to detecting mineralization can be found in [29].

6.2.3   Protein conformation

Proteins are chains of amino acids, each of which has a basic backbone and distinguishing side chain. The properties of a protein are largely determined by its physical shape, which in turn can be characterized by the shape of its backbone. The shape of a protein's backbone can (with some simplification) be completely defined by the angles between each amino acid and the next. These are called the ϕ and ψ angles, and are almost completely determined by the identities of the amino acids making up the backbone. There are 20 different amino acids that make up proteins in humans.

One of the important problems in bioinformatics is predicting the physical structure of the backbone of a protein (its *conformation*) based on the sequence of amino acids that make it up. The Protein Data Bank (PDB) is a repository for the observed conformations of proteins, and contains about 32,000 examples; this is growing by about 5000 examples per year. Given a sequence of amino acids, the PDB can be queried for the known conformation of that sequence, typically in a range of different contexts. The number of examples, obviously, tends to decrease as the length of the sequence increases.

The PDB could be used to predict protein conformations for sequences of arbitrary lengths by breaking the sequences into short but overlapping

lengths, finding the conformations of each of the short segments, and tl
finding conformations that fit together when they overlap. The problem w
this approach is that any given sequence typically has many different con
mations, and it is not usually clear whether the variations in conformat
represent genuinely different possibilities or are artifacts of some kind.

SVD and SDD can be used to address this problem. Suppose that
choose any three amino acids, A, B, and C, and we search the PDB for
occurrences of ABC adjacently in a recorded protein. The mean number
occurrences for chains of length 3 is about 1650 and the mean number
occurrences of chains of length 4 is 100. There are four bond angles defin
the conformation of the chain ABC, a ϕ and a ψ angle between A and B, a
between B and C. Hence for each ABC we obtain a matrix with (typica
1650 rows and 4 columns.

In principle, the rows in the matrix should be of only a few differ
kinds, representing different possible conformations. Apart from regions c
taining turns, most protein sequences are either part of spirals or sheets. H
ever, the actual data from the PDB describes many different conformatio
often with faint clustering, but with large variation.

A useful way to understand the possible conformations of a bond is
Ramachandran plot, a plot of ϕ versus ψ angles. A Ramachandran plot
half a million bond angle pairs is shown in Figure 6.12. From the Figure, i
clear that there are three most common conformations: the region at the t
left corresponding to so-called β sheets, the region below it corresponding
clockwise helices, and the region to the right corresponding to anticlockw
helices. However, it is also clear that there are many, many other possi
conformations.

For each amino acid triple, for example ASP-VAL-ALA, all examples
be extracted from the PDB. The resulting matrix can be decomposed us
SVD. The first few columns of the resulting U matrix reveal the struct
underneath the apparently different conformations, as shown in Figure 6.
Now it is clear that there are at least four major, well-separated clusters, a
perhaps a number of smaller ones. This sequence is typical – clusters
better defined in the U space, suggesting that the best explanation for
variation seen in the PDB is noise.

Although it is easy to see the clusters by inspection, it is hard to au
mate the selection of clusters and the determination of their boundaries. T
is where SDD can be used.

The SVD is truncated at $k = 3$ and the matrices remultiplied to cre
A_3, a cleaned version of the original bond angle matrix. An SDD is then
plied to this matrix to create a hierarchical clustering in which every exemp
can be allocated to a cluster. For each cluster defined by the SDD (or possi
each cluster that is large enough), an ellipse is fitted to the exemplars fr

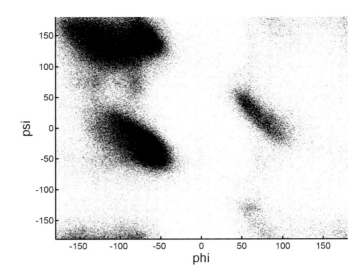

Figure 6.12. *Ramachandran plot of half a million bond angle pa: conformations recorded in the PDB.*

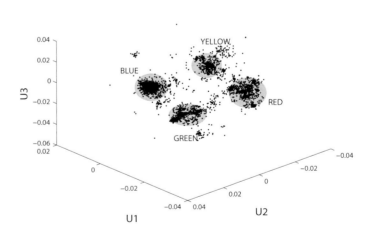

Figure 6.13. *3-dimensional plot of the SVD from the observed bor: angle matrix for ASP-VAL-ALA.*

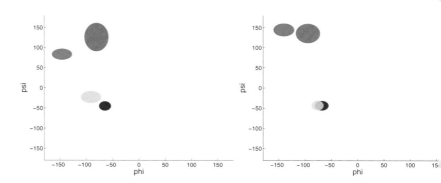

Figure 6.14. *(a) Conformations of the ASP-VAL bond; (b) Con mations of the VAL-ALA bond, from the clusters in Figure 6.13. (See Color Figure 13 in the insert following page 138.)*

that cluster in $\phi - \psi$ space. In other words, each cluster detected by the S produces a cluster in the Ramachandran plot for the first bond and also the second bond. The centroids of these clusters might be called *canon* bond angles since they describe the most likely conformations for a pair amino acids. Figure 6.14 shows the clusters in the two Ramachandran p for the ASP-VAL-ALA sequence. There are four different conformations this sequence: two that involve sheet-like structure, and two α-helices w different spirals.

Once canonical bond angles are known for each pair of amino acids, conformations of longer chains of amino acids can be inferred from the way which these bond angles fit together. This process works well [102]. There some limitations; for example, although the PDB is large, it holds examp only of limited numbers of conformations. For example, at least one seque of length 3 has only four exemplars in the entire PDB.

This application shows how the strengths of SVD and SDD complem each other. SVD is able to remove the noise from the observed protein c formations and display better clusterings of possible conformations. This itself would be useful for investigating a small number of amino acid seque but it is not possible to build an automated clustering for many seque SDD, on the other hand, is able to construct a hierarchical clustering t discovers conformational clusters. By itself, though, it would find many sn clusters. SDD is much more effective because it works with data that l been cleaned up by SVD.

Chapter 7

Independent Component Analysis (ICA)

Independent Component Analysis (ICA) is, in a sense, complementary to singular value decomposition. The factors that an SVD selects are uncorrelated the starting point for ICA is the stronger assumption that the factors are statistically independent. In order for a factorization like ICA to be possible all but one of the distributions of the objects along the axes corresponding to the factors must be non-Gaussian.

ICA was developed for problems such as *blind source separation*. Consider the sounds detected by a set of microphones in a room where a cocktail party is happening. Are these sounds enough to recreate what each person is saying? The fact that people can understand each other at cocktail parties, and can tune out conversations other than the one they are involved in suggests that it *is* possible to do this kind of unmixing. In this situation it is plausible that the different conversations are (more or less) statistically independent, and it turns out that this is critical to the unmixing process.

There are other situations where the values present in a dataset are the result of processes that are independent. We have already mentioned the example of light from a star or galaxy. Here it seems plausible that the processes within the star that generate the light are independent of the processes (for example, gravity) that the light encounters in flight, and that these are independent of atmospheric effects or device effects.

Nevertheless, the assumption of statistical independence deserves some consideration for each dataset. In many real-world settings, processes that seem independent may in fact have some underlying connection that creates a dependence between them. For example, it might seem as if a credit card number should be independent of how much it is used to buy things. A moment's reflection should convince you that this is not true: there are a

sorts of implicit information in a credit card number, including which ba
issued it, which might be correlated with geographical location and so w
economic region; when it was issued, which might be correlated with the po
in the business cycle when it was first issued; and whether it is a single ca
or one of a set billed to the same account.

7.1 Definition

Independent component analysis factors a data matrix, A, whose size is $n \times$
into a product

$$A = C\,F$$

where C is $n \times m$ and F is $m \times m$. The rows of F describe m independent co
ponents and C is a matrix that 'mixes' them to give the observed attribu
of the dataset.

The independent component analysis decomposition of the example matrix

$$
C = \begin{bmatrix}
2.347 & -0.50 & -0.83 & -0.22 & 2.80 & 2.61 & 7.19 & -2.44 \\
1.71 & -0.44 & -0.59 & -0.09 & 3.04 & 0.75 & 7.89 & -2.73 \\
-1.25 & -0.13 & -1.65 & 1.14 & 2.80 & 1.66 & 6.35 & -2.66 \\
1.99 & -0.48 & -2.94 & 2.30 & 4.26 & 0.44 & 7.03 & -3.73 \\
2.52 & -0.48 & -2.84 & -0.04 & 1.47 & 0.39 & 5.74 & -3.33 \\
2.58 & -0.47 & -0.53 & 1.85 & 4.51 & 1.65 & 5.78 & -1.91 \\
1.75 & -0.45 & -2.90 & 0.07 & 3.74 & 2.96 & 4.75 & -2.21 \\
1.84 & -0.47 & -0.75 & 2.08 & 1.60 & 0.46 & 5.12 & -1.27 \\
2.01 & -0.46 & -0.63 & 2.09 & 1.95 & 2.90 & 5.83 & -4.93 \\
2.14 & -0.46 & -2.86 & 2.34 & 1.76 & 2.87 & 7.91 & -1.43 \\
2.10 & 3.03 & -1.65 & 1.15 & 2.79 & 1.67 & 6.36 & -2.66
\end{bmatrix}
$$

$$
F = \begin{bmatrix}
0.84 & -1.17 & 0.77 & -0.87 & 0.53 & -0.48 & 0.12 & -0.18 \\
-0.89 & 0.60 & 0.13 & -0.35 & 1.02 & -0.76 & -0.44 & 1.46 \\
-1.31 & -0.68 & -1.17 & -0.22 & -0.45 & 0.74 & -0.39 & 1.53 \\
0.23 & 0.41 & 0.19 & 0.23 & -0.58 & 0.29 & -1.10 & -1.28 \\
-0.14 & 0.62 & -0.67 & 0.43 & -0.45 & 0.02 & -0.64 & 0.99 \\
-1.42 & -0.40 & -0.78 & 0.03 & -0.01 & 0.56 & 0.10 & 0.15 \\
-0.30 & 0.18 & 0.42 & 0.57 & 0.34 & 0.82 & 1.21 & 1.12 \\
-1.58 & -1.07 & -0.51 & -0.15 & -1.07 & -0.01 & 0.50 & 0.46
\end{bmatrix}
$$

The following are the most obvious differences between a factorizat
such as SVD, and ICA:

- An ICA does not naturally provide a way to reduce the dimensionalit of the data.

- We could always multiply a row of F by some scalar, and divide th corresponding column of C by the same scalar, and still have a decom position. To resolve this ambiguity, the variances of the rows of F ar usually taken to be 1.

- There is no natural ordering of the components, so the rows of F ca be permuted as long as the columns of C are permuted to match. So w cannot truncate the decomposition after a particular number of columr and preserve the most important structure.

  However, there are two possible ways of reordering the matrix explicitl to make sure that the earlier components are more significant than th later ones: order the columns of C in decreasing order of their norm which implies that large mixing coefficients are more important tha small ones; or order the rows of F so that the components whose distr butions are farthest from Gaussian come first. As for SVD, the purpos of these orderings is to focus attention on the most interesting structure of the representation so the one to choose depends on what structure i interesting.

Statistical independence

The difficulty of understanding the difference between two attributes tha are uncorrelated and two attributes that are statistically independent is tha both can be casually described as "knowing the value of one attribute doe not help us to know the value of the other attribute". However, this is tru in two different senses for the two properties.

Two attributes are uncorrelated if

$$E[a_1]E[a_2] = E[a_1 a_2]$$

(where $E[\ ]$ is the expectation) so knowing something about the margina probability of either one reveals nothing about the joint probability. Tw attributes are statistically independent if

$$E[g(a_1)]E[h(a_2)] = E[g(a_1)h(a_2)]$$

that is, even applying non-linear functions g and h to the attributes does nc reveal any correlation information. Statistical independence is obviously stronger property, since it reduces to uncorrelation when g and h are both th identity function.

For example, suppose a_1 and a_2 are constrained to lie on a circle. Th plot of a_1 versus a_2 shows that their values are uncorrelated. However, th squares of their values are very much (negatively) correlated, so they are statistically independent (taking $g(x) = h(x) = x^2$).

Given a data matrix A, we wish to construct C and F so that th product is very close to A. Somewhat surprisingly, the assumption that components are statistically independent is enough to do this. Let us assu that C is invertible – if the components are independent, we would exp that the rows of C would all be quite different. Then we want to define entries of F so that

$$F = C^{-1} A$$

is as independent as possible. So we want the rows F_i and F_j to be und related for $i \neq j$, but we also want $g(F_i)$ and $h(F_j)$ to be uncorrelated suitable choices of g and h. There are a number of ways to choose th functions, based on maximum likelihood, or mutual information. Howev they must all make some use of higher-order statistics of the data or else t decomposition would reduce to SVD. An ICA assumes zero mean data, does SVD, and the data is *whitened* first (essentially performing an SVD) that rows F_i and F_j are already uncorrelated.

The rows of $C^{-1} A$ are linear combinations of rows of A. By the Cent Limit Theorem, sums of non-Gaussian variables are more Gaussian than original variables. Such a linear combination is therefore most non-Gauss when it corresponds to one of the independent components. A strategy finding the independent components, therefore, is to maximize such a lin combination, where the objective function is the uncorrelation of $g(F_i)$ a $h(F_j)$ over all pairs i and j. Each local maximum corresponds to an indep dent component.

An ICA cannot contain more than one Gaussian component, but it contain exactly one. This can sometimes be useful.

ICA is particularly confusing to read about in the literature, since is more natural, in signal processing, to use a matrix transposed from natural orientation of a matrix in data mining.

Normalization

An ICA assumes zero mean data. Most ICA algorithms either require, perform better with, *whitening*, a preprocessing step in which the compone are transformed so that they are uncorrelated; in other words, a step simi to SVD is applied to the dataset before ICA.

It is possible to remove some attributes after whitening, as discussed the SVD section, and then apply ICA to the remaining attributes.

7.2   Interpreting an ICA

An ICA expects that the parts that make up the dataset are statistically inde
pendent and far from Gaussian in shape. The assumption that distribution
are Gaussian tends to be made for most datasets both as a starting point, an
because Gaussians are easy to work with. However, there are clearly appl
cation domains where such simple distributions are not likely and, for these
ICA is the decomposition of choice.

ICA has been applied primarily to applications such as signal processing
and relatives such as financial tracking and biomedical sensing. These domain
have two things in common: it is fairly easy to tell how many component
there are (or should be) in the data; and noise is both expected and we
understood. Other examples of the use of ICA are: analyzing signals from
drilled wells [116]; removing noise from astrophysics data [46]; and chat roor
topic detection [74].

7.2.1   Factor interpretation

The factor interpretation is the natural one for ICA: the rows of F are factor
that are mixed by the entries of C. This is particularly clear in the blin
source separation problem. The rows of A correspond to n signals picked u
by each one of a set of microphones at m time intervals. The F matrix revea
the actual spoken signals at m time intervals, and C shows how these signa
were mixed together to produce those detected by the microphones.

7.2.2   Geometric interpretation

The geometric interpretation treats the rows of C as coordinates in som
geometric space. Plotting some of the columns of C in this way can be
useful way to visualize the structure of the data. Formally, this makes no sens
since the rows of F are not axes (they need not be orthogonal, for example
This means that we cannot apply a metric blindly in this geometric space
Two axes that we are treating as if they are orthogonal might turn out to b
oriented in almost the same direction. In this situation, two points that plo
far apart cannot actually be close together (that is, similar); the problem
that two points that plot close together need not really be as similar as th
plot suggests.

If there are clusters in the data, we expect that such a plot will plac
them along each of the axes, each one corresponding to an independent com
ponent. Hence a three-dimensional visualization of a dataset using ICA wi
always seem like a better clustering than a visualization using SVD. This ha
misled a number of authors to conclude that ICA is inherently better tha
SVD for clustering tasks.

7.2.3   Component interpretation

The component interpretation does not seem to have been used with IC but it is a very natural way to assess the contribution of each independ process to the values in the original dataset. Each row of F describes one the processes that is assumed to be mixed together in the original data. product of the ith column of C with the ith row of F produces a matrix t represents the effect of the ith process on the values of A.

Because of the properties of ICA, each of these outer product matr typically captures a bicluster of objects and attributes. A thresholding al rithm can be applied to an outer product matrix, selecting automatically objects and attributes associated with the bicluster; the threshold determi how strong an association is regarded as significant.

Visualizing the outer product matrix after thresholding can provid view of the internal structure of the bicluster; sorting the outer product ma so that the largest magnitude entries are in the top left-hand corner provi a ranking of the significance of object and/or attributes to the bicluster.

The measures we suggested for finding the most significant structure r not necessarily indicate which outer product is the most interesting. Ot useful measures, such as the number of entries above the threshold, might useful as well.

7.2.4   Graph interpretation

The graph interpretation does not seem helpful for ICA.

7.3   Applying an ICA

7.3.1   Selecting dimensions

As mentioned above, the order of components in an ICA can be made reflect some kind of importance structure. If this is done, then the firs components, for some k, reveal the main structure in the data. Howe there is no principled way to choose k. It may be better to consider e component individually and decide if it has interesting structure.

It is also possible to look at the distribution corresponding to each co ponent. For example, an ICA can contain at most one component tha Gaussian. If such a component is present, it is likely to reflect Gaussian n in the dataset, and so its removal may clean up the data.

The distributions of other components may also reveal that they aris
from particular processes that should be ignored, for example, structure
noise, perhaps related to spatial or temporal artifacts. These may be easy t
identify and remove.

7.3.2 Similarity and clustering

Because of the kinds of applications for which ICA has been used, clusterin
in the 'component', that is attribute domain, is much more often investigate
than clustering in the object domain. However, because the components ar
statistically independent, it is not obvious how to cluster them. The ex
ception is the work by Bach and Jordan [10], who fit the components to
forest-structured graph, deriving the appropriate contrast functions along th
way. Hence, there is an inherent hierarchical cluster on the 'components
This idea cannot be extended to the object domain because it is built-in t
the ICA transform rather than applied as a post-processing step. It is plaus
ble, however, to cluster objects based on their correlation with the clustere
attributes.

Other properties

A technique that is closely related to ICA is LOCOCODE, a neural networ
approach using autoassociators with a particular training technique [53–55
LOCOCODE looks for large, flat basins in weight space and removes thos
weights whose effect is primarily to change position within such a basir
Hence, although the weight space of a neural network is often very larg
LOCOCODE can produce representations that are quite compact. Moreove
the features corresponding to these weights are often clearly 'right' in th
sense of matching the reality of the dataset.

7.4 Algorithm issues

Algorithms and complexity

ICA is really a family of factorizations, parameterized by choices of

- The way in which the deviation from Gaussian is measured – this
  called the *objective* or *contrast* function;

- The algorithm used to find each component, given an objective functior

Objective functions implement the idea of the two nonlinear functi g and h that are used to determine when two components are statistic independent. Objective functions can be divided into two classes, depend on whether they measure the non-Gaussianity and independence of all co ponents at once, or the non-Gaussianity of a component at a time. In first class, some of the objective functions that have been suggested are

- Maximizing the likelihood. However, this requires estimating the pr ability densities of the components, which is complex, and the result function is sensitive to outliers.

- Minimizing the mutual information, using the sum of the differen entropy of each of the variables. Again this is hard to estimate.

In the second class (one component at a time), some of the objective functi that have been suggested are

- Maximizing the negentropy, which is a direct measure of the differe between a given distribution and the Gaussian distribution with same mean and variance. Again this is hard to estimate.

- Maximize higher-order cumulants, such as kurtosis. These are m practical, but take into account mainly the tails of distributions and may be oversensitive to outliers.

- Maximize generalized contrast functions, which are approximations negentropy with particular nice forms and good convergence behav This is the approach taken by the most popular implementation of IC a Matlab package called *FastICA*.

Algorithms for computing independent components are all iterative, dating the mixing matrix and components until they converge. The algorit of choice at present is FastICA, which has good convergence properties a can be used both as a single component at a time, and as a multicompon algorithm. Matlab code is available from `www.cis.hut.fi/projects/i` `fastica/`. The script used for the running example (see Appendix A) ill trates how to use FastICA. In particular, transposes are required to ma the way we orient matrices with the way that FastICA requires them to oriented.

The complexity of ICA depends heavily on the particular objective fu tion and algorithm. For FastICA, convergence is at least quadratic, and see quite fast in practice.

7.5   Applications of ICA

7.5.1   Determining suspicious messages

Governments intercept communications legally, both as part of national security and law enforcement. Increasingly, other organizations are also examining internal communications such as emails, looking for illegal activity or industrial espionage. Especially in the context of national security, a very large number of messages may be intercepted, and only a tiny fraction of these are likely to be of interest. The exact way in which messages are selected for further consideration is secret, but it is known that a watchlist of suspicious words is part of the process [42]. Presumably, messages that use an appropriate number of such suspicious words are treated as suspicious.

Those who would like to communicate without attracting government attention might encrypt their messages. This hides the content, at the expense of making the act of communication more visible. It may be safer to avoid the use of suspicious words, and hide the communication inside the very large amount of routine and innocent communication that is intercepted.

Suspicious words must be replaced by innocent words, but the meaning of each message must still be implicit in the message. This suggests that part of speech will be replaced by the same parts of speech (nouns by nouns, verbs by verbs), and that the same substitution will be used wherever it is needed. So if the word 'bomb' is to be replaced, it will be replaced by another noun, say 'asparagus', whenever it occurs. Such substitutions would, of course, be easy for humans to detect because the semantics of the sentence is altered, but the point of the substitution is to make sure a human never sees the sentence.

The problem now becomes: how is a substituted word to be chosen? A codebook could, of course, be used but this introduces problems with its construction, delivery, and protection which could be difficult for a covert organization. If the substitution is chosen on-the-fly, then the replacement word is likely to differ from the original word in its naturally occurring frequency. For example, 'bomb' is the 3155th most frequent word in English, according to Wordcount (www.wordcount.org), while 'asparagus' is the 27197th most frequent word, so there is a great difference between the two words.

The message with the replacement word may be detectable based on its 'wrong' frequency; all the more so if the same replacement word appears in multiple messages. To look at it another way, conversations involving common words are common; conversations involving rare words are rare. If a common conversation about one topic uses a word or words that would naturally be rare, it begins to look unusual. The converse is also true: a rare conversation is not usually about a common topic. However, this is less important in practice because the tendency is to replace a word by a rarer word – common words

tend to have multiple meanings, which makes it harder for the recipients
work out what meaning was intended. For example, if the original mess
is "the bomb is in position", then it is fairly easy for someone expect
a message about a bomb to understand "the asparagus is in position",
more difficult to be sure about the meaning of "the time is in position".

We illustrate how an ICA can be used to detect a group of messa
that use the same substitutions. The results below are based on an artifi
dataset, a matrix describing 1000 emails, and the frequencies of 400 wo
used in them. The distribution of words in English follows a Zipf distributi
common words are extremely common, with frequencies dropping off quic
so that the frequency of the ith most frequent word is proportional to 1
To model this, the entry in column j for any email is generated by a Pois
distribution with mean $3 * 1/j$, where 3 represents the base mean and
reduces the mean as the column index increases. The resulting matrix
about 4% sparse.

A set of related messages is modelled by inserting 10 extra rows w
the same distribution of entries, but adding a uniformly random block
words in columns 301–306. Each of these messages now contains betw
2 and 3 overlapped words with each other unusual message, and the wo
in these columns occur much more frequently than would be expected fr
the underlying Zipf distribution. We use only columns 201–400, since m
messages contain many common words, so the early columns of the data
tell us little about the relationships among messages.

Figure 7.1 shows a visualization of the first three components of the
matrix of an ICA of this dataset. The set of messages with correlated a
unusual word use is clearly distinct from the other messages.

This detection technique selects only messages that involve unusual wo
use in a correlated way. Sets of messages that have correlated word use, bu
words with typical frequencies (that is, ordinary conversations) do not sh
up as outliers. This is shown in Figure 7.2.

Furthermore, unusual word usage by itself does not cause messages
be selected. Figure 7.3 shows what happens when each of the extra rows u
unusual words, but they are not the same words.

An ICA analysis of such data has exactly the right properties: it dete
conversations whose subjects suggest that conversations about them sho
be rare. On the other hand, it does not detect ordinary conversations abo
ordinary things, nor does it detect unusual word use that is not part o
conversation.

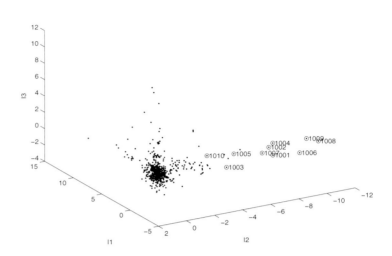

Figure 7.1. *3-dimensional plot from an ICA of messages with co related unusual word use. The messages of interest are circled.*

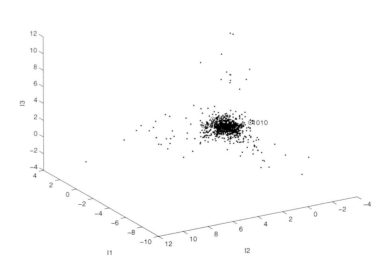

Figure 7.2. *3-dimensional plot from an ICA of messages with co related ordinary word use. The messages of interest are circled.*

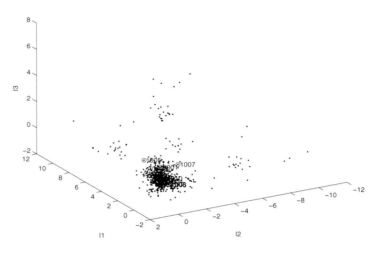

Figure 7.3. *3-dimensional plot from an ICA of messages with usual word use. The messages of interest are circled.*

7.5.2   Removing spatial artifacts from microarrays

We saw in Section 3.5.5 how an SVD can be applied to microarray data select genes that are likely to be of most interest. For some technologies, t channel cDNA microarrays, the spots themselves are printed on each s using technology derived from ordinary printers. Unfortunately, the way which spots are printed creates artifacts that are large enough to call i question the results obtained from such microarrays.

Two-channel arrays print the same amino acid chain repeatedly to each spot. Each sample from a condition class (for example, a patient wit disease) is mixed with a background sample and the combination is allo to hybridize with a slide. The condition and background samples are e labelled with a different marker that fluoresces at different frequencies, a that appear as red and green. When the slide is read, a laser excites each s at each of the two frequencies and the resulting intensities are measured. ratio of red to green intensity is used as an indication of how much express was present for each gene, relative to the background.

Since particular amino acid chains are assigned to positions on the s at random, we would not, in general, expect to see any systematic patter the measured intensity ratios at different positions across the slide. Figures and 7.5 show views of the important red/green intensity ratio of a slide fi the side edge of the slide and from the bottom edge of the slide, respectiv

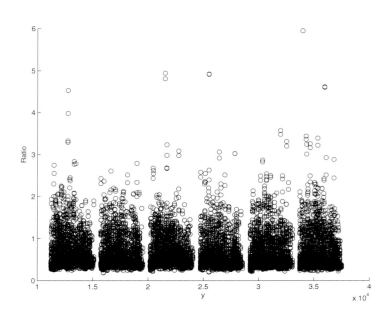

Figure 7.4. *Slide red/green intensity ratio, view from the side.*

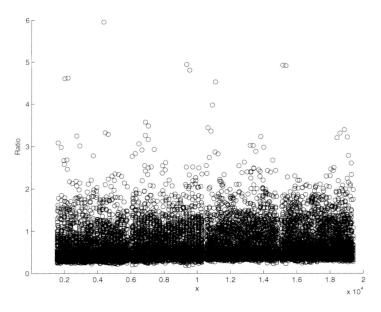

Figure 7.5. *Slide red/green intensity ratio, view from the bottom.*

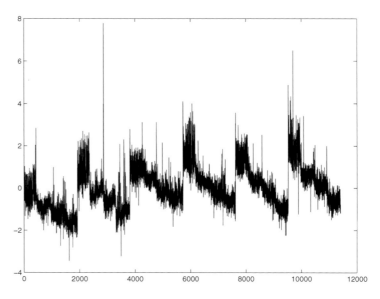

Figure 7.6. *A single component from the slide, with an obv*
spatial artifact related to the printing process.

The fact that the spots are printed in blocks, in six rows of four
very obvious from these figures. However, it is also obvious that there
artifacts in the ratios: both the top and bottom edges of each region in the
show characteristic spatial patterns. These patterns mean that the appar
change in intensity of the expression of some protein between the condi
and background samples depends on where on the slide the spot that t
for that sample is placed. Such spatial artifacts are serious problems
microarray analysis, and they seem to be commonplace. A random sam
of microarray datasets downloaded from the Internet showed problems of
scale in all of them.

ICA can help to remove these spatial artifacts because they appear
single components in the decomposition. For example, for the dataset sh
above, component 10 is shown in the plot in Figure 7.6. It is very clear t
this component captures a 6 × 4 spatial artifact that must be related to
way in which the 24 blocks are printed on the slide. We can see that aver
intensities increase down the slide, and that block intensities decrease ac
the slide, although intensities within each block increase. However, ther
no automatic way to select such components: a human must examine al
the components to see whether any have an obvious spatial component.

Figure 7.7 is another component from the same ICA, showing that th
are spatial artifacts at the edges of each print block, independent of th

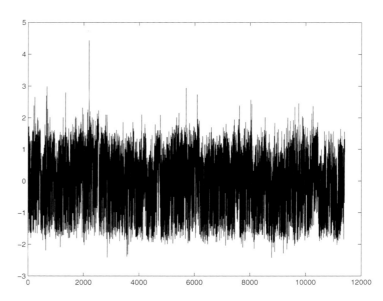

Figure 7.7. *Another component from the slide, with a spatial artifac related to the edges of each printed region.*

related to the print blocks and the slide.

These components can be removed from the decomposition, and th remaining components summed to produce a new version of the dataset, i which the problematic spatial artifacts have been removed. The critical rati of red to green intensity in the resulting dataset is shown in Figures 7.8 an 7.9. There is much less obvious spatial structure in the data, although ther is probably at least one more spatial artifact that could be removed; and th range of values is much less compressed.

ICA's strength in this application is that spatial noise appears as highl non-Gaussian components and so is easily detected and separated by the d composition.

7.5.3 Finding al Qaeda groups

Figure 7.10 shows the C matrix of an ICA applied to a matrix of connection among members of al Qaeda. The dataset matrix is 366×366 and contain a 1 whenever two members of al Qaeda are known to have some kind c relationship, for example carrying out a mission together, being related, c being known friends. The entries in this particular matrix have been sorte into a kind of importance order based on facts known about them, but th

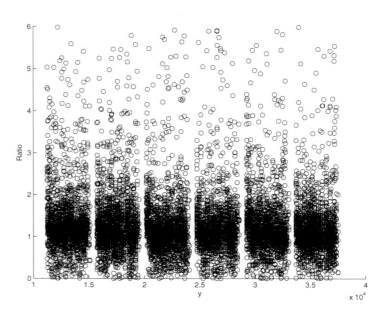

Figure 7.8. *Slide intensity ratio of cleaned data, view from the side.*

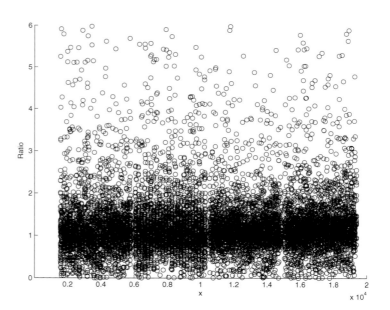

Figure 7.9. *Slide intensity ratio of cleaned data, view from the bottom*

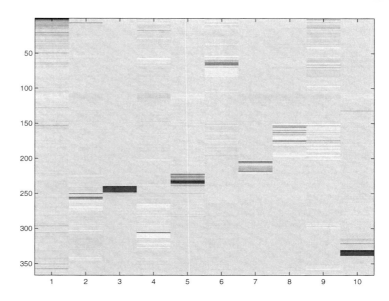

Figure 7.10. *C matrix from an ICA of a matrix of relationship among al Qaeda members. (See also Color Figure 14 in the insert following page 138.)*

examples we show here do not depend on this particular ordering of the rows and columns.

Figure 7.10 clearly shows groups (more or less contiguous as it happens) of members with close connections in each of the components. Component captures the leadership of the organization. Component 2 captures a tight knit group at rows from about 250 to 270 along with an individual at row 13 Other columns show similar groups.

The component matrix, that is the product of column 2 of C and row of F, is shown in Figure 7.11. It is clear that this component captures a grou with a few important members around row 250 and some lesser members i close rows, plus a single important person towards the top of the matri: The important members of this group are Fateh Kamel, Ahmed Ressam, an Mustafa Labsi and the individual representing a connection to the centr: leadership is Mustafa Hamza. These individuals had connections to Group Roubaix in France in the early 1990s and then to the Millennium Bomb Plc in 2000. SVD applied to this same dataset detects the relationship betwee the first three individuals, but is unable to indicate the connection to Hamz because he has connections to many others.

Recall that the rows of the dataset happen to have been arranged i a meaningful order. When they are not, small groups can still be found b

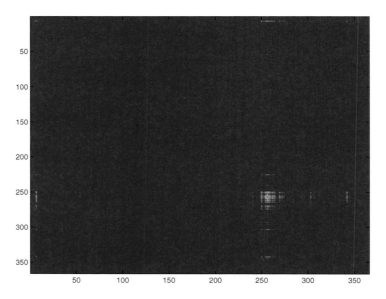

Figure 7.11. *Component matrix of the second component.*

looking for values above a threshold in any of the component matrices.
example, using a threshold of 0.7 of the maximum entry as the cutoff produ
a component matrix where only entries corresponding to Fateh Kamel, Ahr
Ressam, and Mustafa Labsi remain. The matrix can be searched for ent
above the threshold, so that groups can be found algorithmically rather t‌
by inspection.

Notes

Hyvärinen *et al.* [59] attribute ICA to Hérault, Jutten and Ans in the ea
1980s. However, the obvious application of the blind source separation pr‌
lem in sonar, and perhaps radar, suggests that the technique was proba‌
known in military circles well before that. The invention of the FastICA
gorithm by Hyvärinen *et al.* and the wide distribution of Matlab code
greatly increased the use of ICA.

   I am grateful to Marc Sageman for access to the dataset used in the
Qaeda groups example.

Chapter 8

Non-Negative Matrix Factorization (NNMF)

Non-negative matrix factorization is really a class of decompositions whose members are not necessarily closely related to each other. They share the property that they are designed for datasets in which the attribute values are never negative – and it does not make sense for the decomposition matrices to contain negative attribute values either. Such datasets have attributes that count things, or measure quantities, or measure intensities. For example, documents cannot contain negative occurrences of words; images cannot contain negative amounts of each color; chemical reactions cannot involve negative amounts of each reagent, and so on.

A side-effect of this non-negativity property is that the mixing of components that we have seen is one way to understand decompositions can only be *additive*. In other words, a decomposition can only add together components, not subtract them. And the pieces themselves do not have any negative structure, so the combining really is additive – including a new component cannot decrease the size of any matrix entry. It is natural to think of the factors or components as *parts* that are put together additively.

The matrix decompositions we have seen so far will potentially decompose a non-negative matrix in such a way that either the factors or the mixing involve negative values. If there are negative values in the factor matrix, then the factors must somehow describe the absence of something. If there are negative values in the mixing matrix, then constructing the data matrix must require subtracting some components. In the kind of settings mentioned above, neither of these possibilities has a natural interpretation, so the nonnegativity constraint seems appropriate (although it should be kept in mind that imaginary numbers have similar drawbacks, but have turned out to be useful in constructing solutions to a wide variety of problems).

Non-negative matrix factorization (NNMF) was developed to address settings where negative values in the component matrices do not seem app priate. One of the first efficient algorithms that computed an NNMF, due Seung and Lee, also had the property that the decomposition was sparse, t is each entry in the dataset matrix is expressed as the sum of a small num of factors; in other words, the mixing matrix contains many zeroes.

There is a connection between non-negativity and sparsity. Because mixing matrix contains only non-negative entries, an NNMF builds up dataset matrix by adding together factors, which can be thought of as pa making up a whole. Such parts often have a direct physical interpretati In most real-world situations, the way that parts are assembled to mak whole is inherently sparse: most wholes require only a relatively few pa Hence, a parts-based decomposition will tend to be sparse as a side-eff However, it is important to remember that there is no *necessary* link betw non-negativity and sparsity.

8.1   Definition

Unfortunately, most papers in the NNMF literature expect that datasets arranged with the columns representing objects and the rows represent attributes. Such datasets are the transposes of the way in which we have b treating datasets, so once again care is required when reading the literat and using the software packages.

The standard definition for non-negative matrix factorization (NNM of the matrix A is

$$A = W H$$

where A is $m \times n$, W is $m \times r$ and H is $r \times n$, and $r \leq m$. Both W and must contain only non-negative entries. W is the matrix of factors and H the mixing matrix.

However, to be consistent with the other matrix decompositions we h introduced, we will instead define NNMF to be

$$A = C F$$

where A is $n \times m$, C is $n \times r$ and equal to H', and F is $r \times m$ and equal to W So, for us, as usual, C is the mixing matrix and F is the matrix of factors

As usual, r is chosen to be smaller than n or m so that the decomposit is forced to find a compact description of the dataset. A rule of thumb t has been suggested for NNMF in the literature is that

$$r < \frac{nm}{n + m}$$

but it is not at all clear why this is a good choice. The 'generic' data-mining rule of thumb that r should be of the order of $\sqrt{n}$ seems equally plausible.

Unlike the other matrix decompositions we have described, known non negative matrix factorizations are not necessarily closely related. It is no yet clear whether there is some underlying deeper theory to be discovered or whether the existing decompositions are related primarily by the non negativity constraints that they impose. The first fast, simple algorithm for computing an NNMF was described by Seung and Lee [78, 79]. This algorithm tries to minimize

$$||A - WH||_F^2$$

subject to the entries of W and H being non-negative. To avoid degeneracy since $WH = WXX^{-1}H$ for any non-singular X, they constrain the columns of W (rows of F) to sum to unity.

The algorithm is expressed in terms of a pair of update rules that are applied alternately (although there is some empirical evidence that conver gence is faster if they are applied simultaneously). The rules, expressed in terms of W and H are

$$W_{ij} \leftarrow W_{ij} \sum_k \frac{A_{ik}}{(WH)_{ik}} H_{jk}$$

$$W_{ij} \leftarrow \frac{W_{ij}}{\sum_k W_{kj}}$$

$$H_{ij} \leftarrow H_{ij} \sum_k W_{ki} \frac{A_{kj}}{(WH)_{kj}}$$

The W and H matrices are initialized randomly.

Seung and Lee showed that their NNMF produced decompositions that were naturally sparse for two example datasets: one a dataset of faces where the factors corresponded closely to parts of faces (eyes, moustaches, etc.), and the other a word-document dataset, where the factors corresponded fairly well to topics. These decompositions were sparse in both senses: the parts were compact descriptions of real objects, and the observed data could be explained using a relatively small number of parts.

However, the observed sparsity of the decomposition of these two datasets does not seem to generalize to other datasets; that is sparsity is an occasional fortuitous outcome of using their algorithm, but is not guaranteed. Nor is con vergence of their recurrences guaranteed, although this does not seem to be problem in practice.

Several researchers have generalized NNMF by adding explicit terms to the minimization problem to penalize lack of sparsity in W and H. This is slightly problematic since it forces sparsity into datasets that might not

in fact, be naturally sparse. In some settings, it is clear that sparsity
decomposition should be expected, but these sparsifying NNMF algorith
should be used with caution for datasets whose likely sparsity properties
not obvious.

Explicit sparsity NNMF algorithms seek to minimize an objective fu
tion of the form

$$||A - WH||_F^2 + \text{penalty}(W) + \text{penalty}(H)$$

The simplest form of penalty might be the Frobenius norms of W and
For example, Lin [82] defines two projected gradient algorithms that use s
constraints.

Hoyer [56] defines a sparseness constraint for a vector x to be

$$\text{sparseness}(x) = \frac{\sqrt{n} - (\sum |x_i|)/\sqrt{(\sum x_i^2)}}{\sqrt{n} - 1}$$

where n is the length of x. This function takes values between 0 and 1, w
value 0 if all elements are of equal magnitudes, and 1 when there is only
non-zero element, varying smoothly in between. His implementation allc
sparseness values for both W and H to be arguments.

Shahnaz *et al.* [97] define a gradient-descent conjugate least squa
algorithm for NNMF that includes a smoothing parameter.

Dhillon and Sra [34] generalize the problem using Bregman divergenc
a way of describing the minimization problem in terms of a very large clas
difference functions for the difference between A and WH, as well as a la
class of penalty functions. They show that most other definitions of NN]
decompositions are special cases of their Bregman divergence formulation

In general, convergence behavior of NNMF algorithms is not well
derstood, so most algorithms require users to provide a maximum numbe
iterations. It is also possible to terminate when the differences between
and H from one iteration to the next become sufficiently small.

The other parameter that must be set in computing an NNMF is
number of components, r. Unlike previous matrix decompositions, NN]
does not construct one component at a time, so it is not trivial to decide w]
a reasonable choice might be.

The C and F matrices of the NNMF of our example matrix with $r = 8$

shown below.

$$
C = \begin{bmatrix}
1.61 & 2.92 & 1.83 & 12.36 & 13.78 & 1.03 & 1.96 & 0.51 \\
7.17 & 5.94 & 3.37 & 8.40 & 14.11 & 0.90 & 0.60 & 2.51 \\
4.68 & 18.96 & 0.05 & 1.47 & 8.29 & 1.83 & 0.71 & 0.01 \\
15.49 & 8.09 & 0.10 & 0.00 & 0.00 & 5.79 & 1.30 & 13.23 \\
1.65 & 0.03 & 1.17 & 0.06 & 0.48 & 2.00 & 20.61 & 13.99 \\
0.98 & 7.63 & 9.84 & 2.00 & 4.23 & 0.16 & 0.27 & 0.89 \\
0.76 & 7.79 & 0.24 & 0.81 & 5.42 & 0.85 & 7.76 & 4.36 \\
6.61 & 0.50 & 2.85 & 7.44 & 0.00 & 1.16 & 1.72 & 2.72 \\
0.00 & 4.74 & 0.25 & 8.23 & 0.96 & 8.89 & 0.03 & 10.91 \\
4.25 & 1.83 & 0.09 & 13.91 & 1.63 & 2.10 & 11.17 & 0.01 \\
0.86 & 0.21 & 5.29 & 0.10 & 10.29 & 20.12 & 0.05 & 0.09
\end{bmatrix}
$$

$$
F = \begin{bmatrix}
0.18 & 0.16 & 0.26 & 0.15 & 0.00 & 0.01 & 0.16 & 0.07 \\
0.00 & 0.34 & 0.00 & 0.31 & 0.03 & 0.27 & 0.03 & 0.03 \\
0.14 & 0.01 & 0.13 & 0.10 & 0.04 & 0.21 & 0.00 & 0.37 \\
0.00 & 0.00 & 0.12 & 0.17 & 0.09 & 0.38 & 0.20 & 0.04 \\
0.00 & 0.02 & 0.01 & 0.01 & 0.22 & 0.01 & 0.27 & 0.47 \\
0.00 & 0.28 & 0.20 & 0.10 & 0.27 & 0.04 & 0.00 & 0.11 \\
0.11 & 0.09 & 0.26 & 0.11 & 0.16 & 0.01 & 0.27 & 0.00 \\
0.45 & 0.08 & 0.11 & 0.02 & 0.23 & 0.10 & 0.00 & 0.00
\end{bmatrix}
$$

8.2 Interpreting an NNMF

8.2.1 Factor interpretation

The natural way to interpret an NNMF is as defining a set of factors, and mixing of those factors to produce the observed data. Because of the non negativity, the factors can be interpreted as parts, and the mixing as additio of parts. In both ways, NNMF has attractive simplicity. The factor inter pretation has been successful when the underlying data are images or signal However, this is not automatically the case, and the factors produced are no always easy to interpret in the context of other problem domains.

8.2.2 Geometric interpretation

Since the rows of H have no natural interpretation as axes, there is no natura interpretation of NNMF geometrically. Nevertheless, it can be useful to plo the entries of either matrix as if they were coordinates, as we did for ICA. A

before, two points that are located far apart must be dissimilar, but two po
located close together are not necessarily similar. Because of the addi
nature of the model, distance from the origin is important, because the c
way for a point to be far from the origin is either to use parts with la
magnitude entries, or to use large mixing coefficients or both. Conversel
point can be close to the origin only if *both* the entries of its parts are sm
and its mixing coefficients are small.

8.2.3   Component interpretation

Each component is the product of a column of C and a row of F; when
has interesting structure, it is a kind of bicluster since it captures objects
attributes that are related. Because of non-negativity, these biclusters r
be easy to relate to the problem domain. For example, this interpretatio
helpful for *topic detection* in word-document datasets, since word frequen
are inherently positive, and topics are exactly the biclusters in such a data
For some microarrays where the measured values are non-negative, NN
provides an alternative way to find biclusters of genes and conditions.
faces, the biclusters describe facial features, such as moustaches, that app
on some subset of the faces.

*A bar plot of the first layer matrix from the NNMF of our example ma
shows clearly that it captures the block of large values in the lower left cor
of the matrix.*

8.2.4   Graph interpretation

As with the graph interpretation of other decompositions, the graph interp
tation of an NNMF is a tripartite graph, with one set of nodes correspond
to objects, a second set corresponding to the r components, and a third
corresponding to the attributes. The differences in this case are that
graph is sparse because the two decomposition matrices are sparse; and t
the constraints on the weights of edges are all based on sums of non-negat
quantities, so they can be bounded more easily. Graphs whose edges h
only positive weights are also inherently easier to understand.

8.3   Applying an NNMF

8.3.1   Selecting factors

Like ICA, an NNMF does not order components in any particular order, s
is not trivial to select 'interesting' or 'important' factors. When the rows o

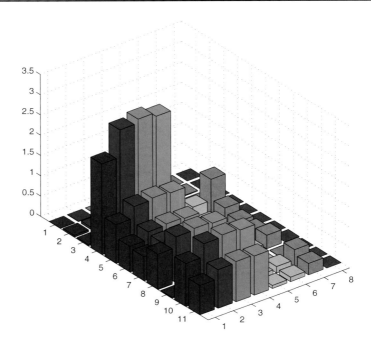

Figure 8.1. *Product of the first column of W and the first row of F from the NNMF of the example matrix.*

are normalized so that the row sums are unity, the norms of the columns of *C* can be used as one way to order the components. These norms represent the extent to which each particular factor plays a role in the description of all the objects. While this is a sensible measure, it is not clear that it capture the real property of interest, since we expect most parts to play a role i representing only a few objects. Another possibility is to use the Frobeniu norm of each component, that is the product of the ith column of C and th ith row of F. This represents the mass of the ith bicluster, which should b meaningful in most contexts.

8.3.2 Denoising

Little attention has been paid to the effects of noise in NNMF, partly becaus many applications start from integral, non-negative data, where noise is eas to see and remove in the raw data. If noise is widespread in the data, tha is most values in the dataset have been slightly altered as the result of nois then it is not clear what happens in the decomposition. Especially for thos algorithms that enforce sparsity, it seems unlikely that noise will appear i one or a few components. Instead, it may be spread throughout the othe meaningful components.

Simple experiments suggest that NNMF is sensitive to noise, especia to small but widely distributed noise, so that the biclustering structure char substantially when only modest Gaussian distributed noise is introduced. T issue needs further research.

8.3.3   Similarity and clustering

The rows of C can be used as the basis of clustering the objects. This wo be expected to perform better than clustering directly on the rows of A b because the rows of C are of length r rather than m, and because the spa fication of both C and F should make it easier to find meaningful bounda within the geometric space containing the rows of C. Similarity is now ba on the similarity of *mixture* coefficients rather than similarity of proper of the objects. Of course, this assumes implicitly that the parts have b properly discovered.

8.4   Algorithm issues

8.4.1   Algorithms and complexity

Because of the relatively large number of algorithms proposed to comp the NNMF, little can be said in general about complexity. Lin [82] makes point that algorithms seem to exist on a continuum, one end of which conta algorithms for which the cost of each step of the minimization is high but wl require few steps; and the other end of which contains algorithms for wl the step cost is small, but which require many steps. Since the entries of and H must be updated in each step, the total complexity is $\Omega(nrs)$ whe is the number of steps. The constants may be large, and memory hierar effects may become significant for large matrices since the access pattern d not have spatial locality.

Matlab code for the Seung and Lee algorithm is given in [78] and r also be downloaded from `journalclub.mit.edu` under "Computational N roscience". Matlab code for Lin's algorithm is available in [82]. Mat code for the algorithm used in Shahnaz *et al.* [97] is described in the per. Hoyer has made a suite of NNMF programs available at his web (**www.cs.helsinki.fi/u/phoyer/**).

8.4.2   Updating

Because the algorithms are iterative, it is straightforward to handle the uation when entries of A change; the algorithm must be run for a few m steps, but convergence will be rapid if the changes are small.

It is also possible to handle the situation where the size of A change by adding rows or columns, since extra rows can be added to W and/or H to match. These rows can be initialized to random values, just as W and H are initialized in the basic algorithm. Further iterations of the algorithm wi provide the updated decomposition.

8.5 Applications of NNMF

8.5.1 Topic detection

Shahnaz *et al.* apply their conjugate gradient constrained least squares algo rithm to topic detection in two corpora, the Reuters data corpus and TDT (the Topic Detection and Tracking Phase 2 corpus). They use the maximu value in each row of C to allocate a document to one of the r topics.

Their results show good agreement with the known topics for small num bers of topics, dropping quickly (for Reuters) and more slowly (for TDT2) a the number of clusters increases. Interestingly, they observe that the perfo mance of NNMF decreases when the bicluster sizes are significantly differen In other words, performance is good when most of the clusters are roughly th same size, but decreases when there are some large clusters and some sma ones.

In these datasets, the actual topics are known. The r outer produc matrices can, in general, be used to generate the topics automatically, base on regions where there are large magnitude values in these matrices.

8.5.2 Microarray analysis

NNMF has also been used to analyze microarrays. Recall that microarray produce data about the expression levels of a large number of genes (oligom cleotides) in samples. This information can be combined into a single arra with rows corresponding to each gene, and columns corresponding to eac sample. In general, we would expect that only some genes correlate with eac possible condition of the samples. This is a biclustering problem.

Carmona-Saez *et al.* have addressed this problem using NNMF extende to enforce smoothness and sparsity. They do this by adding a third matrix t the decomposition

$$A = W S H$$

where S is an $r \times r$ smoothing matrix given by

$$S = (1 - \theta)I + \theta \frac{\mathbf{11}'}{r}$$

where θ is a control parameter between 0 and 1. When θ is close to 1, smoothing matrix contains a value close to $1/r$ everywhere; when θ is c to 0, the smoothing matrix is close to the identity matrix, that is almost of the value is close to the diagonal.

They then minimize
$$||A - WSH||_F^2$$

subject to W and H remaining non-negative. This is similar to the smooth used by Hoyer, without requiring the use of explicit penalty terms.

Carmona-Saez *et al.* [23] report strong results on a number of artifi and real datasets for which the correct results are known with some confiden In particular, the outer product matrices reveal biclusters clearly.

8.5.3   Mineral exploration revisited

Recall that in Section 6.2.2 we looked at the problem of predicting dee covered mineralization based on the partial element concentrations of sur or near-surface samples. We can apply NNMF to this same dataset to ge feel for what kinds of results it can supply.

Figures 8.2, 8.3, and 8.4 show plots of the first three dimensions fr an SVD, from Seung and Lee's basic NNMF algorithm and from the Gradi Descent Conjugate Least Squares NNMF algorithm, using the dataset fr Section 6.2.2. Recall that this dataset has 215 rows and 238 columns, desc ing concentrations from five digestions. For the SVD, the matrix has b moved to the positive orthant by subtracting the smallest value from al the entries (so we expect the first component to capture the global magnit of the data entries).

The most obvious feature of these four plots is how little difference th is among them. Although many points are in slightly different positions, overall structure is quite well preserved among them, and the same samples outliers in all three. This suggests that NNMF is not substantially differ from SVD; some of its apparent clarity may be the result of the kind normalization that is natural in a non-negative setting. Hoyer's C ma has many more zero entries than the other algorithms, but the objects t remain still show some of the same structure as the other matrices, with same outliers.

Figures 8.6, 8.7, 8.8, and 8.9 show the outer product plots for 6 c ponents from an SVD, from the Seung and Lee NNMF, and from the C dient Descent Conjugate Least Squares NNMF. The first point to notic how much the SVD's outer products differ from the NNMF outer produ This clearly illustrates the way in which SVD factors are global descripti whereas the factors of NNMF are local (that is, parts). As we observed earl

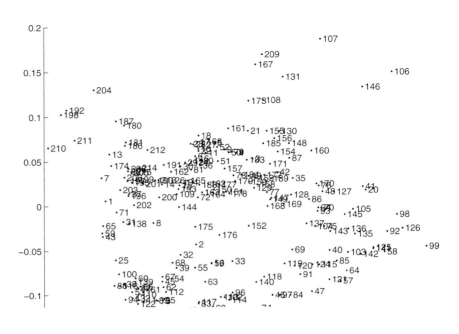

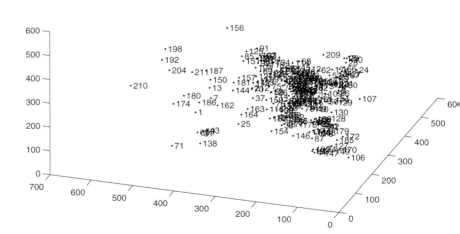

Figure 8.3. *Plot of the C matrix from Seung and Lee's NNMF.*

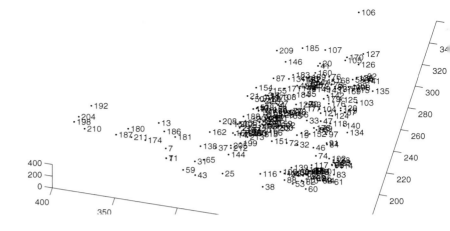

g

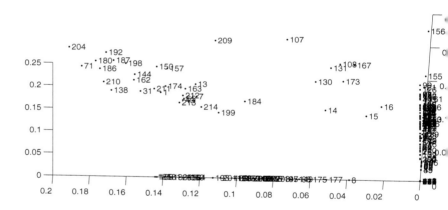

Figure 8.5. *Plot of the C matrix from Hoyer's NNMF.*

the second SVD outer product provides some hint about where the mine
ization might be, but it is not obvious without more domain knowledge.
Seung-Lee NNMF provides strong biclusters, one of which – component
corresponds well to locations and signals of the buried mineralization. Th
are a number of other strong biclusters, several of which do correspond
other interesting geochemistry visible in the dataset. The Gradient Desc
algorithm seems to have applied too strong a sparseness constraint to one

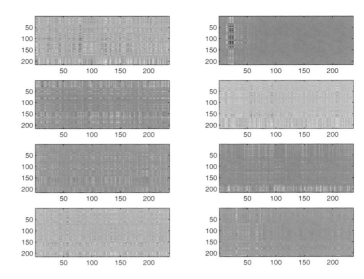

Figure 8.6. *Outer product plots for the SVD. (See also Color Figur 15 in the insert following page 138.)*

the decomposition matrices, so that it suggests clusterings of the digestion but does not show biclusters. Component 3 does reveal the underlying mir eralization, but it would be hard to see this if the answer was not alread known. Hoyer's NNMF shows the effect of strong sparsity in the outer proc ucts. Components 2 and 4 capture the locations of mineralization, but the divide the element concentration signals across two different components which does not appear to have any physical significance. Again, it seems a if too much sparsity can obscure the structure in the data.

Notes

The idea of positive matrix factorization seems to have been originally de veloped by P. Paatero at the University of Helsinki, and to be popular i the computational science community (e.g. [62]). Interest in positive ma trix factorization increased when the fast algorithm for *non-negative matr factorization* (NNMF), based on iterative update, was developed by Lee an Seung [79], particularly as they were able to show that it produced intuitivel reasonable factorizations for a face recognition problem [78]. Donoho an Stodden provided some justification for when NNMF decomposes a datase into parts in [36].

NNMF continues to attract attention because of the natural non-negativ of many application domains and the difficulty of interpreting negative entri

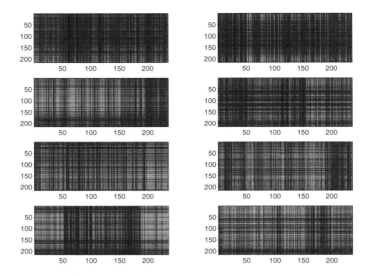

Figure 8.7. *Outer product plots for Seung and Lee's NNMF. (also Color Figure 16 in the insert following page 138.)*

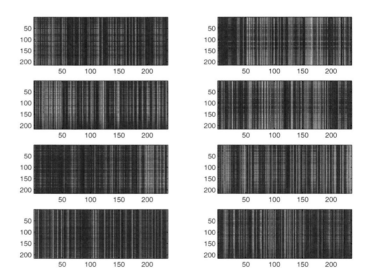

Figure 8.8. *Outer product plots for Gradient Descent Conju Least Squares NNMF. (See also Color Figure 17 in the insert following 138.)*

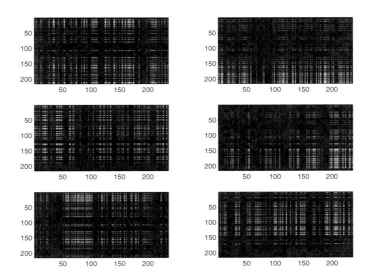

Figure 8.9. *Outer product plots for Hoyer's NNMF. (See also Col Figure 18 in the insert following page 138.)*

in the matrices of other decompositions. However, it is not clear wheth NNMF is a new decomposition, or simply a redefinition of SVD or somethin close to it. Ding, for example, argues for equivalences between NNMF and form of k-means clustering.

Sparsity and its relationship to non-negative decompositions remain problematic. There are clearly datasets where sparsity and non-negativit are natural and complementary. It is less clear that sparsity should always b a goal of a non-negative decomposition.

Chapter 9

Tensors

So far, we have considered only datasets in which there are two kinds of entities: objects and attributes. However, in several settings it is more natural to consider a dataset as having three, or even more, kinds of entities. For example, we have considered word-document matrices derived from emails and seen how to extract relationships between words and documents. Another possibility, though, would be to build a 3-dimensional matrix indexed by words, documents, and times, whose entries would represent the frequency of words used in documents during given time periods.

We can look at such a matrix from three directions. If we look at it from the 'front', then we see a word-document matrix; if we look at it from the 'side', then we see a document-time matrix from which, for example, we can see how email usage changes with time; if we look at it from the 'top', then we see a word-time matrix from which, for example, we can look at trends in word usage over time. All of these are sensible data-mining tasks. However, looking at the three dimensions two at a time means that we lose information about the mutual dependencies among all three attributes. For example, certain words are more popular in emails sent after work, it may be difficult to see this in any of the pairwise slices.

We would like to be able to investigate the structure of such 3-dimensional data in an integrated way, decomposing the matrix directly, rather than slicing or flattening it and then using one of the two-dimensional decompositions. Matrices with extents in three or more dimensions are called tensors. Tensor decompositions allow us to decompose such matrices directly. We will concentrate on tensor decompositions of three-dimensional matrices, but the ideas and techniques extend to matrices of more than three dimensions.

Unfortunately, decompositions get more complicated in two ways: first, notation becomes more cumbersome; second, many of the uniqueness results that hold in the two-dimensional case do not extend to the three-dimensio case.

Tensor decompositions are usually attributed to Ledyard Tucker w in 1966, tried to generalize SVD. The first tensor decomposition we will l at has come to be named after him.

9.1   The Tucker3 tensor decomposition

Suppose that the dataset matrix, A, is $n \times m \times p$. A tensor decomposit expresses A as the combination of four matrices N, M, P, and C. Matri is $n \times nn$, matrix M is $m \times mm$, matrix P is $p \times pp$, and matrix C, whic called the *core* matrix is $nn \times mm \times pp$. The new extents, nn, mm, and are not necessarily bounded by the extents of A.

One of the difficulties of working with tensors is the complexity of writ matrix equations describing the relationships that hold among the matri To avoid the overhead of new notation, we will express these, as much possible, in pointwise form. The Tucker3 tensor decomposition can be writ pointwise as

$$A_{ijk} = \sum_{\alpha=1}^{nn} \sum_{\beta=}^{mm} \sum_{\gamma=1}^{pp} C_{\alpha\beta\gamma} N_{i\alpha} M_{j\beta} P_{k\gamma}$$

Tensor decompositions are calculated by minimizing the difference tween the left and right hand sides of the equation, usually using an alterna least squares algorithm. The matrices N, M, and P are much like the gular vector matrices in an SVD; their rows correspond to the entities in respective dimensions of A, and each column behaves like an eigenvector. called the *core* matrix. Its entries are much like singular values and descr the importance of each triple of component columns in N, M, and P. Tucker3 tensor decomposition is shown in Figure 9.1.

There are many degrees of freedom in a Tucker3 decomposition, so i usual to make the columns of N, M, and P orthogonal. Not only does this l with analysis, but it also makes the computation of the decomposition fas However, when several components may be similar, for example because t are highly correlated, requiring orthogonality may make the decomposit harder to interpret, since these components will be forced to look dissimi It is also usual to normalize the lengths of the columns in the compon matrices to unity.

There are three issues to consider in constructing and interpretin Tucker3 decomposition.

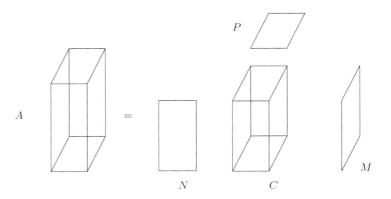

Figure 9.1. *The basic tensor decomposition.*

How to choose the number of components

The 'free' sizes of the component matrices are bounded above by

$$nn \le m \cdot p$$
$$mm \le n \cdot p$$
$$pp \le n \cdot m$$

The 'best' decomposition depends on making good choices for these sizes. The problem is rather like the decision about how many components to retain in an SVD, but getting the equivalent of the scree plot of singular values is more complicated. The search space of possible sizes is large, making exhaustive search unattractive.

The quality of a Tucker3 decomposition is the ratio of the sum of the squares of the elements of the core matrix, C, to the sum of the squares of the elements of A. The larger the elements of C, the more that the tensor decomposition is generalizing the underlying structures, requiring less variability in N, P, and M, and so permitting more agreement in C. Tucker suggested that the values for nn, mm, and pp be approximated by computing eigenvalues and eigenvectors of flattenings of A, that is reducing A to a large two-dimensional matrix by placing, say, its faces side by side.

For example, if we flatten A by taking its planes from the front, we obtain a 2-dimensional matrix $\tilde{A}$ of size $n \times mp$. If we compute the eigendecomposition

$$\tilde{A}\tilde{A}' = N\Lambda N'$$

then we get the matrix N, which might not necessarily be of full rank m. We can then choose some value for nn and keep only the first nn columns

N. Similarly, we can flatten A in the two other possible directions to get and P, and choose mm and pp columns respectively. From these matrices, can compute an approximate core, by first computing G_P of size $nn \times mm$ using $G_P = N'\tilde{A}(P \otimes M)$ where $\otimes$ is the Kronecker product, which multip M by each element of P regarded as a scalar and assembles the result in positional order corresponding to P. Putting the faces of G_P together gi the approximate core. It does not matter which direction is favored in dc this computation, so without loss of generality we have written it in term the faces. The fit for this choice of nn, mm, and pp can be computed fi this approximate core. However, this process needs to be iterated over a la space of choices of the component matrix column sizes.

The approximate fit approach can be used to compute the *largest* j sible approximate core by allowing nn, mm, and pp to take their maxir values, that is use all of the eigenvectors of the respective decompositic Each approximate core matrix for smaller values of nn, mm, and pp can computed by taking subsets of the maximal one. A process called DIF: [68] can now be used to find the best possible values for the three paramet traversing the space of parameters in decreasing order of the parameter su $s = nn + mm + pp$ [68].

As problems become large, even this procedure becomes cumberso For this reason, Tucker3 decompositions have had limited use for data-min problems. A better heuristic for finding reasonable sizes of the compon matrices is needed.

How to interpret the core matrix

The entries in the core matrix demonstrate how the three entities are coup with large values indicating a strong coupling. The square of each entry C is proportional to the amount of variance that the entry explains. Sort the core array so that large entries appear in the top left corner, and sort the component array columns to match means that the component ma ces resemble the singular vectors in SVD, with the most important variat explained by the earlier columns.

Tucker suggested interpreting the core matrix as describing the lat structure in the data, and the component matrices as mixing this struct to give the observed data [111, page 278], an interpretation we have seen several two-dimensional decompositions. It is not easy to see how to m this idea work in practice.

How to interpret the components

Without loss of generality, consider the first component matrix, N. The row of this matrix are the coordinates of points corresponding to the entities i the rows of A. These coordinates lie in an nn-dimensional submanifold of space of dimension $m \times p$.

We could plot these coordinates as if they were based on Euclidea axes, as we did for ICA. In such a plot, separation indicates dissimilarity, bu closeness does not necessarily indicate similarity (because two axes could b oblique and in almost the same direction, so that points along them woul appear close when they are actually not). Also geometric structure such a collinearity is not to be trusted. Nevertheless, as we have seen, such plots ca provide useful visualization.

If we want a more accurate plot, we must construct a set of orthogona axes to use. Even though the columns of M and P are each orthogonal, the are not necessarily orthogonal to each other. We must therefore construct matrix, R, that orthogonalizes the columns of M and P and then apply R^- to N to produce a set of coordinates relative to these axes. In a similar wa the rows of M and P can be plotted and visualized [67].

If the decomposition is computed in so-called principal axes form, so tha the columns of the component matrix N are the unit normalized eigenvecto of A_N (the flattened form of A with respect to the faces), and the sam respectively, for the other dimensions, then scaling the columns so their sun of squares equal the eigenvalues of $G_N G'_N$ gives coordinates with respect t Euclidean axes automatically [111].

Again by analogy with SVD we can apply techniques such as using som leading subset of the columns of a component matrix as a set of loadings, an select the most significant entities from the rows of the matrix accordingl Clustering techniques can also be applied to the rows of each componer matrix to cluster the relevant entities.

9.2   The CP decomposition

A restricted form of the Tucker3 decomposition was independently disco ered by Carroll and Chang, who called it CANDECOMP, and Harshma who called it PARAFACS. We will accordingly call it the CP decompositio although the PARAFACS name has become most common.

In the CP decomposition, the core matrix of a Tucker3 decomposition is superdiagonal matrix, that is it has non-zero entries only on the superdiagon from one corner to the opposite corner, and each of the component matrice

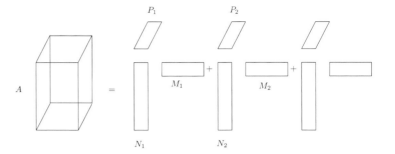

Figure 9.2. *The CP tensor decomposition.*

has the same number of columns. The CP decomposition is therefore a spe
case of the Tucker3 decomposition.

The pointwise description of the CP decomposition is

$$A_{ijk} = \sum_q C_{qqq} N_{iq} M_{jq} P_{kq}$$

The number of terms in the sum is called the *rank* of the tensor, and
no necessary relation to the size of A. The CP decomposition is shown
Figure 9.2.

The CP decomposition is more popular as a data-mining technique t
Tucker3, partly because it is much easier to interpret. The combination of
qth element of N, M, and P is a single outer product whose density struct
makes it a bump detector, and the entry C_{qqq} of the superdiagonal gives
indication of the importance of each outer product.

Like the outer product matrices that we have seen for ICA and NNN
each of the components of the CP decomposition is an outer product mat
ing the shape of the dataset matrix A. Although it may be a little har
in practice, techniques such as visualization can be used to examine the
dimensional outer products. Thresholding of the values can be applied
make it easier to see the structure within the 3-dimensional matrix, and
tries can be sorted so the largest elements are at the top left-hand cor
of the outer product matrix. When the dataset matrix is sorted to mat
indications of the important contributions to each outer product can be fou

9.3   Applications of tensor decompositions

9.3.1   Citation data

Dunlavy *et al.* [40] have used tensors in which each face is an adjace
matrix, but for different kinds of similarity between aspects of a docume

specifically similarity of abstract, of title, of keywords, of authors, and of citations. They use data from eleven SIAM journals and SIAM proceedings for the years 1999-2004, and decompose the resulting $5022 \times 5022 \times 5$ dataset matrix using the CP tensor decomposition.

They interpret the three rank 1 matrices of each component as hub scores, authority scores, and link-type importance scores, generalizing the HITS algorithm of Kleinberg [69]. The largest scores in each of the component matrices identify a kind of community, linked together in multiple ways (that is, via similarities of different kinds). They observe, for example, that components that are similar in the first three link types tend to have similar ties in the strongly weighted words in the hub and authority components; but that this is no longer true for components that are similar in the citation link type because citations are not symmetric. They also develop ways to use the decomposition to compute the centroids of sets of papers. This captures more powerfully the concept of a body of work because it includes papers that are connected to the 'core' papers in more subtle or more complex ways.

9.3.2 Words, documents, and links

Kolda *et al.* [71] apply the CP technique to a dataset whose faces reflect hyperlinks between web pages – entry ij is non-zero if page i links to page – while each face corresponds to a possible word from the anchor text of the link. Anchor text provides a kind of preview of the document to be found at its other end, and so provides useful information beyond that of the link structure. The data was collected by a web crawl and reduced to inter-site rather than inter-page, links.

Their results discover authorities, sites with high scores in a component – but now each component also provides a list of keywords (from the anchor text). Thus each component defines both a set of authorities (or hubs), together with a set of high-scoring words associated with them. This creates ways to extend the HITS algorithm to allow, for example, for sophisticated searches within results.

9.3.3 Users, keywords, and time in chat rooms

Acar *et al.* [2, 3] have applied the Tucker3 decomposition to chatroom data where the three dimensions are users, keywords, and time windows. They argue that the CP decomposition is not appropriate for such data because the number of components should be different in each dimension.

They show that the interaction patterns in the data cannot be fully captured by flattenings of this data, and so there is a genuine advantage to using a tensor decomposition, rather than some two-dimensional decomposition.

In another paper, Acar *et al.* [1] consider the analysis of EEG data locating the focus of epileptic seizures. They show that the Tucker3 decom sition of a matrix whose three dimensions are electrodes, time, and scale encoding of frequency) is superior to SVD and kernel PCA on the pairw slice matrices of electrodes versus time, and electrodes versus frequency.

9.4 Algorithmic issues

An N-way toolbox has been developed by C. A. Andersson and R. Bro [9] Matlab. B. Bader and T.G. Kolda [11] have more recently developed a ten toolbox. Both of these toolboxes implement the decompositions descri here, as well as many basic tensor operations.

The decomposition algorithms can be modified to enforce extra prop ties, for example non-negativity of the entries of the component matrices. many zeroes for the core matrix.

Arbitrary non-singular matrices can be applied to the component r trices, as long as they are compensated for by applying the matching inve transformations to the core matrix. This makes many transformations s' as rotations possible, and there is a considerable literature addressing how apply rotations to achieve structure in the core matrix.

Notes

The Tucker3 tensor decomposition was developed by Tucker [111]. The decomposition was developed independently by Harshman [51] and Car and Chang [24].

Tensor decompositions have many degrees of freedom, so it is possi to impose many constraints to get specialized forms of tensor decompositio Many more or less principled specializations have been developed within social science community, and new specializations are being developed wit the linear algebra and data-mining communities. See, for example, [98].

Chapter 10

Conclusion

Datasets in science, engineering, medicine, social sciences, and derived from the Internet have three significant properties:

- They are extremely large, holding millions or even billions of records partly because collection can often be automated.

  This property implies that practical data mining is limited to technique whose complexity is at worst quadratic in the size of the dataset, an even this may be impractical for the largest datasets. Increasingly, is not the cost of the computation that is the limiting factor, but the cost of moving data up through the memory hierarchy. Practical data mining algorithms must not require more than a constant number of passes through the data, and incremental algorithms are increasingly attractive for datasets that change over time.

- The entries are usually the combination of values that have arisen from a number of processes. This happens in two distinct ways. First, some datasets are collected from the real world using sensing devices, and so the values collected are affected by the variability of the real world, the properties of the sensing devices, and by noise. Second, humans can be thought of as complex multiprocess systems internally, so any data values that depend on humans are already combinations from many contributing activities. For example, humans are inconsistent, even from day to day, in their opinions about almost everything.

  This property implies that it is not useful to analyze datasets as if they represented a single process. Such analysis cannot produce good models because it is based on deeply-flawed assumptions. Instead, analysis must first address the multiprocess character of complex datasets: removing 'noise' in its most general sense where present, and uncovering the lat or component structure. Only then is it sensible to model the data conventional data-mining ways, often component by component.

- Properties of each individual object arise from the values of only so of its attributes, while its other attributes are more or less irreleva However, the important attributes differ from object to object. In ot words, entries in the matrix are correlated with other entries in lo rather than global, ways.

  This implies two things. First, that the goal of clustering should usu be considered as biclustering; and second, that attribute selection it is usually discussed, cannot help to elucidate structure effectiv Biclustering is already an important research topic; we suggest tha will become the standard form of clustering for complex datasets.

We have suggested that *matrix decompositions* are ideal tools to att the problems of modelling such large and complex datasets. Matrix decom sitions are all, underneath, forms of Expectation-Maximization that assu particular forms of the underlying distributions. However, implementati of EM typically allow only a limited number of relatively simple distributi to be used, so this connection is not very useful for practical data mining. the other hand, each matrix decomposition looks, superficially, to be differ from the others and requires understanding different tools for computing decomposition and different ways of interpreting the results.

Different matrix decompositions are good at exposing different kind structure in a dataset. However, the following list summarizes some of possibilities.

- *Denoising*: This is the simplest use of a matrix decomposition, in wh one or more component is judged to represent noise, and removed fi the dataset. The choice of such components must always be made diciously and with regard to the goals of the data-mining process, si what is noise in one context may be signal in another.

  Since the most common shape for noise is Gaussian, SVD is rigl considered the standard way to remove noise. As we have seen, h ever, ICA can play a useful role both in removing Gaussian noise, a especially in removing structured noise, for example spatial artifact.

- *Finding submanifolds*: When data is collected, it is almost never p sible to guess, in advance, which attributes will turn out to be m significant for the modelling goal. It is therefore common to collect ues for many attributes that are either irrelevant, or related to e other in complex, interlocking ways. In some situations, it is also r ural to collect the data in terms of a large number of attributes;

example, a molecule may be described in terms of the three-dimensional positions of its atoms although there are clearly not as many degrees of freedom in its configuration as this implies. As a result, the 'interesting' structure in a dataset is a low-dimensional manifold hidden within a much higher dimensional space. Matrix decompositions can discover such lower-dimensional structure much more effectively than most attribute-selection techniques, partly because matrix decompositions work from global structure in the data.

- *Finding components*: As discussed above, the critical problem with complex datasets is that they represent the superposition or combination of data values that arise from multiple processes. Matrix decompositions all separate the given dataset into a form where the putative components can be examined. Each component may be modelled separately, or components can be discarded, upweighted, or downweighted, and a modified form of the dataset reconstructed.

  The resulting dataset(s) can then be passed to other data-mining techniques, which can now be relied on to model them effectively since they represent a known process, or they can be clustered directly based on the matrix decomposition.

- *Visualization*: Humans are extremely good at seeing structure. Most matrix decompositions have the following two properties: they can order the components so that the most important or significant ones can be determined; and they provide a sensible way to plot the data values. Together these mean that matrix decompositions make useful visualizations possible. Visualizations can confirm the results of more-technical analysis, and may sometimes suggest structure that would be hard to detect from the models alone.

- *Graph-based clustering*: There are many ways for objects to be similar or dissimilar, but the most powerful ones are those that depend not on the properties of each object alone, but on the context of each pair of objects. Modelling such context is inherently difficult because it may require, for a single pair of objects, considering the entire rest of the dataset. Graph-based clustering provides several ways of reducing this problem to a more straightforward geometrical one, by embedding graphs in Euclidean space in a clever way. Matrix decompositions provide a way to do this approximately in a way that avoids the exponential analysis required to compute many interesting properties, for example good cuts exactly.

- *Extensions to greater numbers of related entities*: Basic matrix decompositions relate objects and attributes. However, as we have seen, there are increasingly common settings where the connections between more than two kinds of entities need to be explored. The multidimensional rays, or tensors, that represent such connections can also be decompo into matrices that reveal some of the underlying relationships, altho this is still a relatively new approach in mainstream data mining.

Each decomposition has its own special strengths. SVD is the s plest decomposition and makes the weakest assumptions about the expec structure in the data – assuming, based on the Law of Large Numbers, t Gaussian distributions are most likely. The biggest advantage of SVD is ability to rank objects or attributes in such a way that both the commor and rarest are ranked low, leaving the most interesting, in a useful sense the top of the ranking.

The strength of SDD is its ability to define a ternary hierarchical cl tering of the objects or attributes. This is especially useful when used al with SVD, since it provides a way to label clusters automatically.

ICA and NNMF are both good at finding components corresponding biclusters, that is components that involve a relatively small number of b objects and attributes. ICA looks for components characterized by stati cal independence, which we saw is a stronger property than uncorrelati NNMF looks for components in the positive orthant. Both decompositi are coming to be appreciated in wider contexts as the problems of find biclusters become more important in applications such as text mining, a mining the results of high-throughput biomedical devices.

The situation where the dataset defines pairwise similarities or affini between entities is completely different from the implicit geometric situat of the standard decompositions. Analysis of such datasets is much easie they can be transformed into a geometric representation, where the stand analyses can be done. This is non-trivial. Obviously, the embedding i geometric space must respect, as far as possible, the pairwise affinities. H ever, it must also place non-adjacent points in such a way that their dista apart in the geometry corresponds in a sensible way to their separation in graph. There are a number of ways to define separation in the graph, a each leads to a different embedding. Many of these definitions are motiva by the correspondence between simple boundaries in the geometric space, a high-quality cuts in the graph. Analysis of graph data is still relatively n and these analysis techniques will surely become more important.

Finally, tensors enable us to extend ideas of relationships among two ferent kinds of entities to simultaneous relations among three or more differ kinds of entities. A great deal of work is being done on the technologies of t sor decomposition, but tensor decompositions have not been applied to ma data-mining tasks (except, of course, for long-standing applications in fie such as chemometrics). Much of the development of tensor decomposit was done in social sciences, working with three-dimensional but quite sm datasets. The development of new algorithms and interpretation technique may lead to renewed applications to larger datasets in the social sciences.

Appendix A

Matlab Scripts to generate example matrix decompositions

The scripts here are designed to show you how to get started calculating you own matrix decompositions. Matlab makes it easy to do many things in way that are clever but hard to read. These scripts do things in the most obviou ways, to make them easy to understand. With some Matlab experience, yo will be able to improve them.

This is the main script. It is unfortunately somewhat cluttered with th output machinery that generates the files that are used in the main part c the book.

```
function base(mfn)
format short;

% machinery to create result files
% make directory name
dmfn = ['figs' mfn];
status = mkdir(dmfn);

a = csvread([mfn '.csv']);

n = size(a,1)
m = size(a,2)

ifile = [dmfn '/initial.txt'];
inf = fopen(ifile,'w');
for i = 1:n
fprintf(inf,'%6.0f &',a(i,:));
fprintf(inf,'\\\\ \n');
```

```
end
fclose(inf);

% normalize a

na = zeros(n,m);
for j = 1:m
if std(a(:,j)) == 0
na(:,j) = (a(:,j) - mean(a(:,j)));
else
na(:,j) = (a(:,j) - mean(a(:,j)))./std(a(:,j));
end;
end;

% svd of unnormalized data

[u,s,v] = svd(a,0);

ufile = [dmfn '/' mfn '.u'];
vfile = [dmfn '/' mfn '.v'];
sfile = [dmfn '/' mfn '.s'];
psfile = [dmfn '/' mfn ];
uf = fopen(ufile,'w');
for i = 1:size(u,1)
fprintf(uf,'%6.2f &',u(i,:));
fprintf(uf,'\\\\ \n');
end
fclose(uf);

vf = fopen(vfile,'w');
for i = 1:size(v,1)
fprintf(vf,'%6.2f &',v(i,:));
fprintf(vf,'\\\\ \n');
end
fclose(vf);

sf = fopen(sfile,'w');
for i = 1:size(v,1)
fprintf(sf,'%6.2f &',s(i,:));
fprintf(sf,'\\\\ \n');
end
fclose(sf);

ss = diag(s);
```

```
ssvf1 = fopen([dmfn '/singvalue1.txt'],'w');
fprintf(ssvf1,'%4.2f, followed by %4.2f and %4.2f.\n',ss(1),ss(
fclose(ssvf1);

% labelled objects
figure;

plot3(u(:,1),u(:,2),u(:,3),'r.','MarkerSize',12)
for i = 1:size(u,1)
   text(u(i,1), u(i,2), u(i,3), [' ' int2str(i)],'FontSize',11)
end
view([-150,30])
axis('auto')
xlabel('U1')
ylabel('U2')
zlabel('U3')

print('-deps2', [dmfn '/u.eps']);

% labelled attributes
figure

hold on;
plot3(v(:,1),v(:,2),v(:,3),'r.','MarkerSize',12);
for i=1:size(v,1)
text(v(i,1),v(i,2),v(i,3),[' ' int2str(i)],'FontSize',11);
end
view([-150,30])
axis('auto')
xlabel('V1')
ylabel('V2')
zlabel('V3')

print('-deps2', [dmfn '/v.eps']);

% singular values
figure;

plot(ss,'-k+','MarkerSize',10)
ylabel('s')

print('-deps2', [dmfn '/s.eps']);

% US and SV
us = u * s;
```

```
vs = v * s;

%  plot u and v matrices scaled by the singular values
figure;

plot3(us(:,1),us(:,2),us(:,3),'r.','MarkerSize',12)
for i = 1:size(u,1)
   text(us(i,1), us(i,2), us(i,3), [' ' int2str(i)],'FontSize
end
view([-150,30])
axis('auto')
xlabel('US1')
ylabel('US2')
zlabel('US3')

print('-deps2', [dmfn '/us.eps']);

% labelled attributes
figure

hold on;
plot3(vs(:,1),vs(:,2),vs(:,3),'r.','MarkerSize',12);
for i=1:size(v,1)
text(vs(i,1),vs(i,2),vs(i,3),[' ' int2str(i)],'FontSize',11);
end
view([-150,30])
axis('auto')
xlabel('VS1')
ylabel('VS2')
zlabel('VS3')

print('-deps2', [dmfn '/sv.eps']);

% repeat for normalized A

[u,s,v] = svd(na,0);

nufile = [dmfn '/' mfn '.nu'];
nvfile = [dmfn '/' mfn '.nv'];
nsfile = [dmfn '/' mfn '.ns'];

uf = fopen(nufile,'w');
for i = 1:size(u,1)
fprintf(uf,'%6.2f &',u(i,:));
fprintf(uf,'\\\\ \n');
```

```
end
fclose(uf);

vf = fopen(nvfile,'w');
for i = 1:size(v,1)
fprintf(vf,'%6.2f &',v(i,:));
fprintf(vf,'\\\\ \n');
end
fclose(vf);

sf = fopen(nsfile,'w');
for i = 1:size(v,1)
fprintf(sf,'%6.2f &',s(i,:));
fprintf(sf,'\\\\ \n');
end
fclose(sf);

ss = diag(s);

ssvf2 = fopen([dmfn '/singvalue2.txt'],'w');
fprintf(ssvf2,'%4.2f, followed by %4.2f and %4.2f.\n',ss(1),ss(
fclose(ssvf2);

% calculate entropy

sssum = 0;
for i = 1:m
sssum = sssum + ss(i).*ss(i);
end

for i = 1:m
f(i) = ss(i).*ss(i)/sssum;
end

entrsum = 0;
for i = 1:m
entrsum = entrsum + f(i) * log(f(i));
end
entropy = -entrsum/(log(m))

ef = fopen([dmfn '/entropy.txt'],'w');
fprintf(ef,'%5.3f, %5.3f, %5.3f, %5.3f, %5.3f, %5.3f, %7.5f, an
fprintf(ef,' The entropy for this dataset is %5.3f,', entropy);
fclose(ef);
```

```
% labelled objects (normalized)
figure;

plot3(u(:,1),u(:,2),u(:,3),'r.','MarkerSize',12)
for i = 1:size(u,1)
   text(u(i,1), u(i,2), u(i,3), [' ' int2str(i)],'FontSize',1
end
view([-150,30])
axis('auto')
xlabel('U1')
ylabel('U2')
zlabel('U3')

print('-deps2', [dmfn '/nu.eps']);

% labelled attributes (normalized)
figure

hold on;
plot3(v(:,1),v(:,2),v(:,3),'r.','MarkerSize',12);
for i=1:size(v,1)
text(v(i,1),v(i,2),v(i,3),[' ' int2str(i)],'FontSize',11);
end
view([-150,30])
axis('auto')
xlabel('V1')
ylabel('V2')
zlabel('V3')

print('-deps2', [dmfn '/nv.eps']);

% singular values

figure;

plot(ss,'-k+','MarkerSize',10)
ylabel('s')

print('-deps2', [dmfn '/ns.eps']);

% matrix with extra rows

aa = [a; 1 1 1 1 1 1 1 1; 9 9 9 9 9 9 9 9];

[u,s,v] = svd(aa,0);
```

```
aaufile = [dmfn '/' mfn '.aau'];
aavfile = [dmfn '/' mfn '.aav'];
aasfile = [dmfn '/' mfn '.aas'];

aauf = fopen(aaufile,'w');
for i = 1:size(u,1)
fprintf(aauf,'%6.2f &',u(i,:));
fprintf(aauf,'\\\\ \n');
end
fclose(aauf);

aavf = fopen(aavfile,'w');
for i = 1:size(v,1)
fprintf(aavf,'%6.2f &',u(i,:));
fprintf(aavf,'\\\\ \n');
end
fclose(aavf);

aasf = fopen(aasfile,'w');
for i = 1:size(v,1)
fprintf(aasf,'%6.2f &',s(i,:));
fprintf(aasf,'\\\\ \n');
end
fclose(aasf);

% labelled objects
figure;

plot3(u(:,1),u(:,2),u(:,3),'r.','MarkerSize',12)
for i = 1:size(u,1)
    text(u(i,1), u(i,2), u(i,3), [' ' int2str(i)],'FontSize',11)
end
view([-160,30])
axis('auto')
xlabel('U1')
ylabel('U2')
zlabel('U3')

print('-deps2', [dmfn '/orient.eps']);

% matrix with extra rows 2

ab = [a; 1 1 1 1 9 9 9 9; 9 9 9 9 1 1 1 1];
```

```
[u,s,v] = svd(ab,0);

abufile = [dmfn '/' mfn '.abu'];
abvfile = [dmfn '/' mfn '.abv'];
absfile = [dmfn '/' mfn '.abs'];
abuf = fopen(abufile,'w');
for i = 1:size(u,1)
fprintf(abuf,'%6.2f &',u(i,:));
fprintf(abuf,'\\\\ \n');
end
fclose(abuf);

abvf = fopen(abvfile,'w');
for i = 1:size(v,1)
fprintf(abvf,'%6.2f &',v(i,:));
fprintf(abvf,'\\\\ \n');
end
fclose(abvf);

absf = fopen(absfile,'w');
for i = 1:size(v,1)
fprintf(absf,'%6.2f &',s(i,:));
fprintf(absf,'\\\\ \n');
end
fclose(absf);

% labelled objects
figure;

plot3(u(:,1),u(:,2),u(:,3),'r.','MarkerSize',12)
for i = 1:size(u,1)
   text(u(i,1), u(i,2), u(i,3), [' ' int2str(i)],'FontSize',1
end
view([-100,30])
axis('auto')
xlabel('U1')
ylabel('U2')
zlabel('U3')

print('-deps2', [dmfn '/nextorient.eps']);

% output A1 A2

[u,s,v] = svd(na,0);
```

```
a1 = u(:,1) * s(1,1) * v(:,1)';
a2 = u(:,2) * s(2,2) * v(:,2)';
a1file = [dmfn '/a1.txt'];
a2file = [dmfn '/a2.txt'];
a3file = [dmfn '/a3.txt'];
autoafile = [dmfn '/autoa.txt'];

a1f = fopen(a1file,'w');
for i = 1:size(u,1)
fprintf(a1f,'%6.2f &',a1(i,:));
fprintf(a1f,'\\\\ \n');
end
fclose(a1f);

a2f = fopen(a2file,'w');
for i = 1:size(u,1)
fprintf(a2f,'%6.2f &',a2(i,:));
fprintf(a2f,'\\\\ \n');
end
fclose(a2f);

asum = a1 + a2;

a3f = fopen(a3file,'w');
for i = 1:size(u,1)
fprintf(a3f,'%6.2f &',asum(i,:));
fprintf(a3f,'\\\\ \n');
end
fclose(a3f);

autoa = asum * asum';

autoaf = fopen(autoafile,'w');
for i = 1:size(u,1)
fprintf(autoaf,'%6.2f &',autoa(i,:));
fprintf(autoaf,'\\\\ \n');
end
fclose(autoaf);

% end of svd section

k = 8;

[d,x,y] = sdd(na,k);
```

```
yfile = [dmfn '/' mfn '.y'];
dfile = [dmfn '/' mfn '.d'];
xfile = [dmfn '/' mfn '.x'];

xf = fopen(xfile,'w');
for i = 1:size(x,1)
fprintf(xf,'%6.0f &',x(i,:));
fprintf(xf,'\\\\ \n');
end
fclose(xf);

yf = fopen(yfile,'w');
for i = 1:size(y,1)
fprintf(yf,'%6.0f &',y(i,:));
fprintf(yf,'\\\\ \n');
end
fclose(yf);

df = fopen(dfile,'w');
for i = 1:size(y,1)
fprintf(df,'%6.2f &',d(i,:));
fprintf(df,'\\\\ \n');
end
fclose(df);

op = x(:,2) * y(:,2)'

opf = fopen([dmfn '/op1.txt'],'w');
for i = 1:size(x,1)
fprintf(opf,'%4.0f &',op(i,:));
fprintf(opf,'\\\\ \n');
end
fclose(opf);


% sdd with largest volume components moved to the front

[d,x,y] = smsdd(na,k);

yfile = [dmfn '/' mfn '.sy'];
dfile = [dmfn '/' mfn '.sd'];
xfile = [dmfn '/' mfn '.sx'];

xf = fopen(xfile,'w');
for i = 1:size(x,1)
```

```
fprintf(xf,'%6.0f &',x(i,:));
fprintf(xf,'\\\\ \n');
end
fclose(xf);


yf = fopen(yfile,'w');
for i = 1:size(y,1)
fprintf(yf,'%6.0f &',y(i,:));
fprintf(yf,'\\\\ \n');
end
fclose(yf);


df = fopen(dfile,'w');
for i = 1:size(y,1)
fprintf(df,'%6.2f &',d(i,:));
fprintf(df,'\\\\ \n');
end
fclose(df);


op = x(:,2) * y(:,2)'

opf = fopen([dmfn '/op2.txt'],'w');
for i = 1:size(x,1)
fprintf(opf,'%4.0f &',op(i,:));
fprintf(opf,'\\\\ \n');
end
fclose(opf);


n = size(na,1);
m = size(na,2);

% plots of regions

figure;
for r = 1:k
peak = d(r) * x(:,r) * y(:,r)';
nn = norm(peak,'fro');
bar3(peak,1.0,'detached','y');
title(['Bump at level ' int2str(r)]);
ylabel('Objects');
xlabel('Attributes');
zlabel('Bump direction');
print('-deps2',[dmfn '/peak' int2str(r)]);
print('-depsc2',[dmfn '/cpeak' int2str(r)]);
clf;
```

```
end

% 3 level sdd for objects

figure;
view([-150,30])
axis('auto')
xlabel('U1')
ylabel('U2')
zlabel('U3')
legend;
sz = n;

xdis;

print('-deps2',[dmfn '/sso'] );
print('-depsc2',[dmfn '/csso'] );

% 3 level sdd for attributes

figure;

view([-150,30])
axis('auto')
xlabel('V1')
ylabel('V2')
zlabel('V3')
sz = m;

ydis;

print('-deps2',[dmfn '/ssa'] );
print('-depsc2',[dmfn '/cssa'] );

%  nnmf using Seung and Lee code

% transpose because algorithm expects attributes as rows
A = a';

[n m] = size(A);
r = 8; % choose rank for the factorization
maxiter = 300; % choose the maximum number of iterations

W = rand(n,r); % randomly initialize basis
W = W./(ones(n,1)*sum(W)); % normalize column sums
```

```
H = rand(r,m); % randomly initialize encodings
eps = 1e-9; % set your own tolerance

for iter=1:maxiter
H = H.*(W'*((A+eps)./(W*H+eps)));
W = W.*(((A+eps)./(W*H+eps))*H');
W = W./(ones(n,1)*sum(W));
end

C = H';
F = W';

cf = fopen([dmfn '/c.txt'],'w');
for i = 1:size(C,1)
fprintf(cf,'%6.2f &',C(i,:));
fprintf(cf,'\\\\ \n');
end
fclose(cf);

ff = fopen([dmfn '/f.txt'],'w');
for i = 1:size(F,1)
fprintf(ff,'%6.2f &',F(i,:));
fprintf(ff,'\\\\ \n');
end
fclose(ff);
```

These are display routines for the SDD labelling of objects and at-tributes, respectively.

Objects:

```
hold on

for i = 1:sz
split1 = x(i,1);
switch split1
   case 1
   split2 = x(i,2);
   switch split2
      case 1,
      split3 = x(i,3);
      switch split3
        case 1,
        plot3(u(i,1),u(i,2),u(i,3),'r.','ButtonDownFcn',num2str
```

```
text(u(i,1),u(i,2),u(i,3),['    ' int2str(i) ],'FontSize',11);
      case 0,
      plot3(u(i,1),u(i,2),u(i,3),'k.','ButtonDownFcn',num2s
text(u(i,1),u(i,2),u(i,3),['    ' int2str(i) ],'FontSize',11);
      case -1,
      plot3(u(i,1),u(i,2),u(i,3),'b.','ButtonDownFcn',num2s
text(u(i,1),u(i,2),u(i,3),['    ' int2str(i) ],'FontSize',11);
      end
      case 0,
      split3 = x(i,3);
      switch split3,
        case 1,
        plot3(u(i,1),u(i,2),u(i,3),'ro','ButtonDownFcn',num2s
text(u(i,1),u(i,2),u(i,3),['    ' int2str(i) ],'FontSize',11);
        case 0,
        plot3(u(i,1),u(i,2),u(i,3),'ko','ButtonDownFcn',num2s
text(u(i,1),u(i,2),u(i,3),['    ' int2str(i) ],'FontSize',11);
        case -1,
        plot3(u(i,1),u(i,2),u(i,3),'bo','ButtonDownFcn',num2s
text(u(i,1),u(i,2),u(i,3),['    ' int2str(i) ],'FontSize',11);
      end
      case -1,
      split3 = x(i,3);
      switch split3
        case 1,
        plot3(u(i,1),u(i,2),u(i,3),'rx','ButtonDownFcn',num2s
text(u(i,1),u(i,2),u(i,3),['    ' int2str(i) ],'FontSize',11);
        case 0,
        plot3(u(i,1),u(i,2),u(i,3),'kx','ButtonDownFcn',num2s
text(u(i,1),u(i,2),u(i,3),['    ' int2str(i) ],'FontSize',11);
        case -1,
        plot3(u(i,1),u(i,2),u(i,3),'bx','ButtonDownFcn',num2s
text(u(i,1),u(i,2),u(i,3),['    ' int2str(i) ],'FontSize',11);
      end
    end
    case 0
    split2 = x(i,2);
    switch split2
      case 1,
      split3 = x(i,3);
      switch split3
        case 1,
        plot3(u(i,1),u(i,2),u(i,3),'r+','ButtonDownFcn',num2s
text(u(i,1),u(i,2),u(i,3),['    ' int2str(i) ],'FontSize',11);
        case 0,
```

```
        plot3(u(i,1),u(i,2),u(i,3),'k+','ButtonDownFcn',num2str
text(u(i,1),u(i,2),u(i,3),['  ' int2str(i) ],'FontSize',11);
        case -1,
        plot3(u(i,1),u(i,2),u(i,3),'b+','ButtonDownFcn',num2str
text(u(i,1),u(i,2),u(i,3),['  ' int2str(i) ],'FontSize',11);
      end
      case 0,
      split3 = x(i,3);
      switch split3,
        case 1,
        plot3(u(i,1),u(i,2),u(i,3),'r*','ButtonDownFcn',num2str
text(u(i,1),u(i,2),u(i,3),['  ' int2str(i) ],'FontSize',11);
        case 0,
        plot3(u(i,1),u(i,2),u(i,3),'k*','ButtonDownFcn',num2str
text(u(i,1),u(i,2),u(i,3),['  ' int2str(i) ],'FontSize',11);
        case -1,
        plot3(u(i,1),u(i,2),u(i,3),'b*','ButtonDownFcn',num2str
text(u(i,1),u(i,2),u(i,3),['  ' int2str(i) ],'FontSize',11);
      end
      case -1,
      split3 = x(i,3);
      switch split3
        case 1,
        plot3(u(i,1),u(i,2),u(i,3),'rs','ButtonDownFcn',num2str
text(u(i,1),u(i,2),u(i,3),['  ' int2str(i) ],'FontSize',11);
        case 0,
        plot3(u(i,1),u(i,2),u(i,3),'ks','ButtonDownFcn',num2str
text(u(i,1),u(i,2),u(i,3),['  ' int2str(i) ],'FontSize',11);
        case -1,
        plot3(u(i,1),u(i,2),u(i,3),'bs','ButtonDownFcn',num2str
text(u(i,1),u(i,2),u(i,3),['  ' int2str(i) ],'FontSize',11);
      end
    end
    case -1
    split2 = x(i,2);
    switch split2
      case 1,
      split3 = x(i,3);
      switch split3
        case 1,
        plot3(u(i,1),u(i,2),u(i,3),'rd','ButtonDownFcn',num2str
text(u(i,1),u(i,2),u(i,3),['  ' int2str(i) ],'FontSize',11);
        case 0,
        plot3(u(i,1),u(i,2),u(i,3),'kd','ButtonDownFcn',num2str
text(u(i,1),u(i,2),u(i,3),['  ' int2str(i) ],'FontSize',11);
```

```
        case -1,
        plot3(u(i,1),u(i,2),u(i,3),'bd','ButtonDownFcn',num2s
text(u(i,1),u(i,2),u(i,3),['  ' int2str(i) ],'FontSize',11);
      end
      case 0,
      split3 = x(i,3);
      switch split3,
        case 1,
        plot3(u(i,1),u(i,2),u(i,3),'rv','ButtonDownFcn',num2s
text(u(i,1),u(i,2),u(i,3),['  ' int2str(i) ],'FontSize',11);
        case 0,
        plot3(u(i,1),u(i,2),u(i,3),'kv','ButtonDownFcn',num2s
text(u(i,1),u(i,2),u(i,3),['  ' int2str(i) ],'FontSize',11);
        case -1,
        plot3(u(i,1),u(i,2),u(i,3),'bv','ButtonDownFcn',num2s
text(u(i,1),u(i,2),u(i,3),['  ' int2str(i) ],'FontSize',11);
      end
      case -1,
      split3 = x(i,3);
      switch split3
        case 1,
        plot3(u(i,1),u(i,2),u(i,3),'r^','ButtonDownFcn',num2s
text(u(i,1),u(i,2),u(i,3),['  ' int2str(i) ],'FontSize',11);
        case 0,
        plot3(u(i,1),u(i,2),u(i,3),'k^','ButtonDownFcn',num2s
text(u(i,1),u(i,2),u(i,3),['  ' int2str(i) ],'FontSize',11);
        case -1,
        plot3(u(i,1),u(i,2),u(i,3),'b^','ButtonDownFcn',num2s
text(u(i,1),u(i,2),u(i,3),['  ' int2str(i) ],'FontSize',11);
      end
    end
end
end


Attributes:


hold on

for i = 1:sz
split1 = y(i,1);
switch split1
   case 1
   split2 = y(i,2);
```

```
    switch split2
       case 1,
       split3 = y(i,3);
       switch split3
         case 1,
         plot3(v(i,1),v(i,2),v(i,3),'r.','ButtonDownFcn',num2str
text(v(i,1),v(i,2),v(i,3),['   ' int2str(i) ],'FontSize',11);
         case 0,
         plot3(v(i,1),v(i,2),v(i,3),'k.','ButtonDownFcn',num2str
text(v(i,1),v(i,2),v(i,3),['   ' int2str(i) ],'FontSize',11);
         case -1,
         plot3(v(i,1),v(i,2),v(i,3),'b.','ButtonDownFcn',num2str
text(v(i,1),v(i,2),v(i,3),['   ' int2str(i) ],'FontSize',11);
       end
       case 0,
       split3 = y(i,3);
       switch split3,
         case 1,
         plot3(v(i,1),v(i,2),v(i,3),'ro','ButtonDownFcn',num2str
text(v(i,1),v(i,2),v(i,3),['   ' int2str(i) ],'FontSize',11);
         case 0,
         plot3(v(i,1),v(i,2),v(i,3),'ko','ButtonDownFcn',num2str
text(v(i,1),v(i,2),v(i,3),['   ' int2str(i) ],'FontSize',11);
         case -1,
         plot3(v(i,1),v(i,2),v(i,3),'bo','ButtonDownFcn',num2str
text(v(i,1),v(i,2),v(i,3),['   ' int2str(i) ],'FontSize',11);
       end
       case -1,
       split3 = y(i,3);
       switch split3
         case 1,
         plot3(v(i,1),v(i,2),v(i,3),'rx','ButtonDownFcn',num2str
text(v(i,1),v(i,2),v(i,3),['   ' int2str(i) ],'FontSize',11);
         case 0,
         plot3(v(i,1),v(i,2),v(i,3),'kx','ButtonDownFcn',num2str
text(v(i,1),v(i,2),v(i,3),['   ' int2str(i) ],'FontSize',11);
         case -1,
         plot3(v(i,1),v(i,2),v(i,3),'bx','ButtonDownFcn',num2str
text(v(i,1),v(i,2),v(i,3),['   ' int2str(i) ],'FontSize',11);
       end
    end
    case 0
    split2 = y(i,2);
    switch split2
       case 1,
```

```
        split3 = y(i,3);
        switch split3
          case 1,
          plot3(v(i,1),v(i,2),v(i,3),'r+','ButtonDownFcn',num2s
text(v(i,1),v(i,2),v(i,3),['   ' int2str(i) ],'FontSize',11);
          case 0,
          plot3(v(i,1),v(i,2),v(i,3),'k+','ButtonDownFcn',num2s
text(v(i,1),v(i,2),v(i,3),['   ' int2str(i) ],'FontSize',11);
          case -1,
          plot3(v(i,1),v(i,2),v(i,3),'b+','ButtonDownFcn',num2s
text(v(i,1),v(i,2),v(i,3),['   ' int2str(i) ],'FontSize',11);
        end
        case 0,
        split3 = y(i,3);
        switch split3,
          case 1,
          plot3(v(i,1),v(i,2),v(i,3),'r*','ButtonDownFcn',num2s
text(v(i,1),v(i,2),v(i,3),['   ' int2str(i) ],'FontSize',11);
          case 0,
          plot3(v(i,1),v(i,2),v(i,3),'k*','ButtonDownFcn',num2s
text(v(i,1),v(i,2),v(i,3),['   ' int2str(i) ],'FontSize',11);
          case -1,
          plot3(v(i,1),v(i,2),v(i,3),'b*','ButtonDownFcn',num2s
text(v(i,1),v(i,2),v(i,3),['   ' int2str(i) ],'FontSize',11);
        end
        case -1,
        split3 = y(i,3);
        switch split3
          case 1,
          plot3(v(i,1),v(i,2),v(i,3),'rs','ButtonDownFcn',num2s
text(v(i,1),v(i,2),v(i,3),['   ' int2str(i) ],'FontSize',11);
          case 0,
          plot3(v(i,1),v(i,2),v(i,3),'ks','ButtonDownFcn',num2s
text(v(i,1),v(i,2),v(i,3),['   ' int2str(i) ],'FontSize',11);
          case -1,
          plot3(v(i,1),v(i,2),v(i,3),'bs','ButtonDownFcn',num2s
text(v(i,1),v(i,2),v(i,3),['   ' int2str(i) ],'FontSize',11);
        end
    end
    case -1
    split2 = y(i,2);
    switch split2
        case 1,
        split3 = y(i,3);
        switch split3
```

```
        case 1,
        plot3(v(i,1),v(i,2),v(i,3),'rd','ButtonDownFcn',num2str
text(v(i,1),v(i,2),v(i,3),['  ' int2str(i) ],'FontSize',11);
        case 0,
        plot3(v(i,1),v(i,2),v(i,3),'kd','ButtonDownFcn',num2str
text(v(i,1),v(i,2),v(i,3),['  ' int2str(i) ],'FontSize',11);
        case -1,
        plot3(v(i,1),v(i,2),v(i,3),'bd','ButtonDownFcn',num2str
text(v(i,1),v(i,2),v(i,3),['  ' int2str(i) ],'FontSize',11);
      end
      case 0,
      split3 = y(i,3);
      switch split3,
        case 1,
        plot3(v(i,1),v(i,2),v(i,3),'rv','ButtonDownFcn',num2str
text(v(i,1),v(i,2),v(i,3),['  ' int2str(i) ],'FontSize',11);
        case 0,
        plot3(v(i,1),v(i,2),v(i,3),'kv','ButtonDownFcn',num2str
text(v(i,1),v(i,2),v(i,3),['  ' int2str(i) ],'FontSize',11);
        case -1,
        plot3(v(i,1),v(i,2),v(i,3),'bv','ButtonDownFcn',num2str
text(v(i,1),v(i,2),v(i,3),['  ' int2str(i) ],'FontSize',11);
      end
      case -1,
      split3 = y(i,3);
      switch split3
        case 1,
        plot3(v(i,1),v(i,2),v(i,3),'r^','ButtonDownFcn',num2str
text(v(i,1),v(i,2),v(i,3),['  ' int2str(i) ],'FontSize',11);
        case 0,
        plot3(v(i,1),v(i,2),v(i,3),'k^','ButtonDownFcn',num2str
text(v(i,1),v(i,2),v(i,3),['  ' int2str(i) ],'FontSize',11);
        case -1,
        plot3(v(i,1),v(i,2),v(i,3),'b^','ButtonDownFcn',num2str
text(v(i,1),v(i,2),v(i,3),['  ' int2str(i) ],'FontSize',11);
      end
    end
end
end
```

Bibliography

[1] E. Acar, C.A. Bingöl, H. Bingöl, and B. Yener. Computational analys of epileptic focus localization. In *Proceedings of the 24th IASTED In ternational Multi-Conference, Biomedical Engineering*, February 2006

[2] E. Acar, S.A. Çamtepe, M.S. Krishnamoorthy, and B. Yener. Mod elling and multiway analysis of chatroom tensors. In *IEEE Internation Conference on Intelligence and Security Informatics (ISI 2005)*, page 256–268. Springer LNCS 3495, 2005.

[3] E. Acar, S.A. Çamtepe, and B. Yener. Collective sampling and analys of high order tensors for chatroom communication. In *IEEE Intern tional Conference on Intelligence and Security Informatics (ISI2006* pages 213–224. Springer LNCS 3975, 2006.

[4] D. Achlioptas and F. McSherry. Fast computation of low rank matri approximations. In *STOC: ACM Symposium on Theory of Computir (STOC)*, 2001.

[5] R. Agrawal, T. Imielinski, and A.N. Swami. Mining association rul between sets of items in large databases. In P. Buneman and S. Jajodi editors, *Proceedings of the 1993 ACM SIGMOD International Confe ence on Management of Data*, pages 207–216, Washington, D.C., 199°

[6] R. Agrawal and R. Srikant. Fast algorithms for mining association rule In J.B. Bocca, M. Jarke, and C. Zaniolo, editors, *Proceedings of the 20 International Conference on Very Large Data Bases, VLDB*, pages 487 499. Morgan Kaufmann, 1994.

[7] C.J. Alpert and S.-Z. Yao. Spectral partitioning: The more eigenvector the better. In *32nd ACM/IEEE Design Automation Conference*, page 195–200, June 1995.

[8] O. Alter, P.O. Brown, and D. Botstein. Singular value decompositic for genome-wide expression data processing and modeling. *Proceedin of the National Academy of Science*, 97(18):10101–10106, 2000.

[9] C. A. Andersson and R. Bro. The N-way toolbox for MATLAB. *Che* *metrics and Intelligent Laboratory Systems*, 52(1):1–4, 2000.

[10] F.R. Bach and M.I. Jordan. Finding clusters in Independent Com nent Analysis. Technical Report UCB/CSD-02-1209, Computer Scie Division, University of California, Berkeley, 2002.

[11] B.W. Bader and T.G. Kolda. MATLAB tensor classes for fast a rithm prototyping. Technical Report SAND2004-5187, Sandia Natic Laboratories, October 2004.

[12] M. Belkin and P. Niyogi. Laplacian eigenmaps and spectral technic for embedding and clustering. In T. G. Dietterich, S. Becker, Z. Ghahramani, editors, *Advances in Neural Information Proces. Systems 14*, Cambridge, MA, 2002. MIT Press.

[13] M.W. Berry, S.T. Dumais, and G.W. O'Brien. Using linear algebra intelligent information retrieval. *SIAM Review*, 37(4):573–595, 1995

[14] D.L. Boley. Principal direction divisive partitioning. *Data Mining Knowledge Discovery*, 2(4):325–344, 1998.

[15] B.E. Boser, I.M. Guyon, and V.N. Vapnik. A training algorithm optimal margin classifiers. In D. Haussler, editor, *5th Annual A Workshop on COLT*, pages 144–152, Pittsburgh, 1992.

[16] M. Brand. A random walks perspective on maximizing satisfaction profit. In *SIAM International Conference on Data Mining*, pages 12- 2005.

[17] L. Breiman. Random forests–random features. Technical Report Department of Statistics, University of California, Berkeley, Septem 1999.

[18] L. Breiman, J.H. Friedman, R.A. Olshen, and C.J. Stone. *Classifica: and Regression Trees*. Chapman and Hall, New York, 1984.

[19] S. Brin and L. Page. The Anatomy of a Large Scale Hypertextual \ Search Engine. In *Proceedings of the Seventh International Confere of the World Wide Web 7*, pages 107–117, Brisbane, Australia, 199€

[20] S. Brin, L. Page, R. Motwani, and T.Winograd. The PageRank C tion Ranking: Bringing Order to the Web. *Stanford Digital Libra Working Paper*, 1998.

[21] K. Bryan and T. Leise. The $25,000,000,000 eigenvector: The lir algebra behind Google. *SIAM Review*, 48(3):569–581, 2006.

[22] C.J.C. Burges. A tutorial on support vector machines for pattern rec nition. *Data Mining and Knowledge Discovery*, 2:121–167, 1998.

[23] P. Carmona-Saez, R.D. Pascual-Marqui, F. Tirado, J.M. Carazo, an A. Pascual-Montano. Biclustering of gene expression data by nor smooth non-negative matrix factorization. *BMC Bioinformatics*, 7(78 February 2006.

[24] J.D. Carroll and J.-J. Chang. Analysis of individual differences in mu tidimensional scaling via an N-way generalization of "Eckart-Young decomposition. *Psychometrika*, 35:283–319, 1970.

[25] M.T. Chu. On the statistical meaning of truncated singular value de composition. Preprint, 2001.

[26] M.T. Chu, R.E. Funderlic, and G.H. Golub. A rank-one reductio formula and its applications to matrix factorizations. *SIAM Revier* 37:512–530, 1995.

[27] F.R.K. Chung. *Spectral Graph Theory*. Number 92 in CBMS Re gional Conference Series in Mathematics. American Mathematical Sc ciety, 1997.

[28] F.R.K. Chung. Lectures on spectral graph theory. www.math.ucsd.edu ~fan/research/revised.html, 2006.

[29] D.R. Cohen, D.B. Skillicorn, S.Gatehouse, and I. Dalrymple. Signatur detection in geochemical data using singular value decomposition an semi-discrete decomposition. In *21st International Geochemical Exple ration Symposium (IGES)*, Dublin, August 2003.

[30] C. Cortes and V. Vapnik. Support-vector networks. *Machine Learnin* 20, 1995.

[31] N. Cristianini and J. Shawe-Taylor. *An Introduction to Support Vecte Machines and other kernel-based learning methods*. Cambridge Unive: sity Press, 2000.

[32] S. C. Deerwester, S. T. Dumais, T. K. Landauer, G. W. Furnas, an R. A. Harshman. Indexing by latent semantic analysis. *Journal of tł American Society of Information Science*, 41(6):391–407, 1990.

[33] A.P. Dempster, N.M. Laird, and D.B. Rubin. Maximum likelihood fror incomplete data via the EM algorithm. *Journal of the Royal Statistic Society, Series B*, 39:138, 1977.

[34] I.S. Dhillon and S. Sra. Generalized nonnegative matrix approximatior with Bregman divergences. Technical report, University of Texas Austin Department of Computer Sciences, 2005.

[35] C.H.Q. Ding, X. He, and H. Zha. A spectral method to separate discor nected and nearly-disconnected Web graph components. In *KDD 200* pages 275–280, 2001.

[36] D. Donoho and V. Stodden. When does non-negative matrix fac
    ization give a correct decomposition in parts? In *Advances in Ne*
    Information Processing Systems (NIPS) 17, 2004.

[37] H. Drucker, C.J.C. Burges, L. Kaufman, A. Smola, and V. Vapı
    Support vector regression machines. In *Advances in Neural Informaı*
    Processing Systems 9, NIPS, pages 155–161, 1996.

[38] S.T. Dumais, T.A. Letsche, M.L. Littman, and T.K. Landauer.
    tomatic cross-language retrieval using Latent Semantic Indexing.
    AAAI-97 Spring Symposium Series: Cross-Language Text and Spe
    Retrieval, pages 18–24, 1997.

[39] M. Dunham. *Data Mining Introductory and Advanced Topics*. Pren
    Hall, 2003.

[40] D.M. Dunlavy, T.G. Kolda, and W.P. Kegelmeyer. Multilinear alge
    for analyzing data with multiple linkages. Technical Report SAND2(
    2079, Sandia National Laboratories, April 2006.

[41] M. Ester, H.-P. Kriegel, J. Sander, and X. Xu. A density-based a
    rithm for discovering clusters in large spatial databases with noise.
    2nd International Conference on Knowledge Discovery and Data Mir
    (KDD'96), Portland, Oregon, 1996. AAAI Press.

[42] European Parliament Temporary Committee on the ECHELON In
    ception System. Final report on the existence of a global system for
    interception of private and commercial communications (ECHELON
    terception system), 2001.

[43] F. Fouss, A. Pirotte, J.-M. Renders, and M. Saerens. Random-w
    computation of similarities between nodes of a graph, with applicaı
    to collaborative recommendation. *IEEE Transactions on Knowle*
    and Data Engineering, 2006.

[44] J. Friedman and N. Fisher. Bump hunting on high-dimensional dɛ
    Statistics and Computation, 1997.

[45] A. Frieze, R. Kannan, and S. Vempala. Fast Monte-Carlo algorithms
    finding low-rank approximations. In *FOCS '98*, pages 370–378, 199ξ

[46] M. Funaro, E. Oja, and H. Valpola. Artefact detection in astrophys
    image data using independent component analysis. In *Proceedings*
    International Workshop on Independent Component Analysis and B.
    Source Separation, pages 43–48, December 2001.

[47] M. Gladwell. The science of the sleeper. *The New Yorker*, pages 48–
    October 4, 1999.

[48] G.H. Golub and C.F. van Loan. *Matrix Computations*. Johns Hopkin University Press, 3rd edition, 1996.

[49] S.M. Hamilton. Electrochemical mass transport in overburden: A ne model to account for the formation of selective leach geochemical anoma lies in glacial terrain. *Journal of Geochemical Exploration*, pages 155 172, 1998.

[50] D.J. Hand, H. Mannila, and P. Smyth. *Principles of Data Mining*. MI Press, 2000.

[51] R.A. Harshman. Foundations of the PARAFAC procedure: Models an conditions for an "explanatory" multi-modal factor analysis. UCL Working Papers in Phonetics, 16, 1970.

[52] B. Hendrickson. Latent Semantic Analysis and Fiedler retrieval. *Linec Algebra and Applications*, 421:345–355, 2007.

[53] S. Hochreiter and J. Schmidhuber. LOCOCODE versus PCA and ICA In *Proceedings ICANN'98*, pages 669–674, 1998.

[54] S. Hochreiter and J. Schmidhuber. Feature extraction through LC COCODE. *Neural Computation*, 11(3):679–714, 1999.

[55] S. Hochreiter and J. Schmidhuber. Lococode performs nonlinear IC without knowing the number of sources. In *Proceedings of the ICA'9* pages 149–154, 1999.

[56] P.O. Hoyer. Non-negative matrix factorization with sparseness cor straints. *Journal of Machine Learning*, 5:1457–1469, 2004.

[57] L. Hubert, J. Meulman, and W. Heiser. Two purposes for matrix fa torization: A historical appraisal. *SIAM Review*, 42(1):68–82, 2000.

[58] A. Hyvärinen. Survey on independent component analysis. *Neural Con puting Surveys*, 2:94–128, 1999.

[59] A. Hyvärinen, J. Karhunen, and E. Oja. *Independent Component Ana ysis*. John Wiley, 2001.

[60] A. Hyvärinen and E. Oja. Independent component analysis: Algorithn and applications. *Neural Networks*, 13(4–5):411–430, 2000.

[61] S.C. Johnson. Hierarchical clustering schemes. *Psychometrika*, 2:241 254, 1967.

[62] M. Juvela, K. Lehtinen, and P. Paatero. The use of positive matrix fa torization in the analysis of molecular line spectra from the Thumbpri Nebula. In *Proceedings of the Fourth Haystack Conference "Cloud cores and low mass stars,"* Astronomical Society of the Pacific Confe ence Series, volume 65, pages 176–180, 1994.

[63] D. Kalman. A singularly valuable decomposition: The SVD of a mat
     College Math Journal, 27(1), January 1996.

[64] M. Kantardzic. *Data Mining: Concepts, Models, Methods, and A
     rithms*. Wiley-IEEE Press, 2002.

[65] A. Karol and M.-A. Williams. Understanding human strategies
     change: an empirical study. In *TARK '05: Proceedings of the 1*
     conference on Theoretical aspects of rationality and knowledge, pa
     137–149. National University of Singapore, 2005.

[66] P.J. Kennedy, S.J. Simoff, D.B. Skillicorn, and D. Catchpoole. Extra
     ing and explaining biological knowledge in microarray data. In *Pac*
     Asia Knowledge Discovery and Data Mining Conference (PAKDD20
     Sydney, May 2004.

[67] H.A.L. Kiers. Some procedures for displaying results from three-v
     methods. *Journal of Chemometrics*, 14:151–170, 2000.

[68] H.A.L. Kiers and A. der Kinderen. A fast method for choosing the n
     bers of components in Tucker3 analysis. *British Journal of Mathemat*
     and Statistical Psychology, 56:119–125, 2003.

[69] J.M. Kleinberg. Authoritative sources in a hyperlinked environme
     Journal of the ACM, 46(5):604–632, 1999.

[70] G. Kolda and D.P. O'Leary. A semi-discrete matrix decomposition
     latent semantic indexing in information retrieval. *ACM Transacti*
     on Information Systems, 16:322–346, 1998.

[71] T.G. Kolda, B.W. Bader, and J.P. Kenny. Higher-order web link anal
     using multilinear algebra. In *Fifth IEEE International Conference*
     Data Mining, pages 242–249, November 2005.

[72] T.G. Kolda and D.P. O'Leary. Computation and uses of the semidisc
     matrix decomposition. *ACM Transactions on Information Process*
     1999.

[73] T.G. Kolda and D.P. O'Leary. *Latent Semantic Indexing Via A Se*
     Discrete Matrix Decomposition, volume 107 of *IMA Volumes in Ma*
     matics and Its Applications, pages 73–80. Springer Verlag, 1999.

[74] T. Kolenda, L. Hansen, and J. Larsen. Signal detection using IC
     Application to chat room topic spotting. In *Proceedings of ICA '20*
     December 2001.

[75] A. Kontostathis and W.M. Pottenger. Detecting patterns in the
     term-term matrix. Technical Report LU-CSE-02-010, Department
     Computer Science and Engineering, Lehigh University, 2002.

[76] A. Kontostathis and W.M. Pottenger. Improving retrieval performanc with positive and negative equivalence classes of terms. Technical Re port LU-CSE-02-009, Department of Computer Science and Engineer ing, Lehigh University, 2002.

[77] M. Koyutürk and A. Grama. Binary non-orthogonal de composition: A tool for analyzing binary-attributed datasets www.cs.purdue.edu/homes/ayg/RECENT/bnd.ps, 2002.

[78] D.D. Lee and H.S. Seung. Learning the parts of objects by non-negativ matrix factorization. *Nature*, 401:788–791, 1999.

[79] D.D. Lee and H.S. Seung. Algorithms for non-negative matrix factoriza tion. In *NIPS, Neural Information Processing Systems*, pages 556–56. 2000.

[80] M.S. Lewicki and T.J. Sejnowski. Learning overcomplete representa tions. *Neural Computation*, 12(2):337–365, 2000.

[81] D. Liben-Nowell and J. Kleinberg. The link prediction problem fc social networks. In *Proceedings of the Twelfth International Conferenc on Information and Knowledge Management*, pages 556–559, 2003.

[82] C.-J. Lin. Projected gradient methods for non-negative matrix facto ization. *Neural Computation*, to appear.

[83] J.B. MacQueen. Some methods for classification and analysis of mu tivariate observations. In *Proceedings of 5th Berkeley Symposium o Mathematical Statistics and Probability*, volume 1, pages 281–297. Un versity of California Press, 1967.

[84] V.A.J. Maller. Criminal investigation systems: The growing dependenc on advanced computer systems. *Computing and Control Engineerir Journal*, pages 93–100, April 1996.

[85] S. McConnell and D.B. Skillicorn. Semidiscrete decomposition: A bum hunting technique. In *Australasian Data Mining Workshop*, pages 7. 82, December 2002.

[86] M. Meilă and J. Shi. A random walks view of spectral segmentation. I *AI and Statistics (AISTATS)*, 2001.

[87] B. N. Miller, I. Albert, S. K. Lam, J. A. Konstan, and J. Riedl. Movi Lens unplugged: Experiences with an occasionally connected recon mender system. In *IUI'03: Proc. 8th International Conference on I. telligent User Interfaces*, pages 263–266, Miami, Florida, USA, 200 ACM Press.

[88] J. C. Nash. *Compact Numerical Methods for Computers: Linear Algeb and Function Minimisation*. John Wiley & Sons, 1979.

[89] A. Y. Ng, A. X. Zheng, and M. I. Jordan. Link analysis, eigen\
tors and stability. In *Proceedings of the Seventeenth International J*\
Conference on Artificial Intelligence (IJCAI-01), pages 903–910, 20

[90] D.P. O'Leary and S. Peleg. Digital image compression by outer prod\
expansion. *IEEE Transactions on Communications*, 31:441–444, 19

[91] K. Popper. *The Logic of Scientific Discovery*. Hutchinson, Lonc\
1959.

[92] J.R. Quinlan. Induction of decision trees. *Machine Learning*, 1:81–1\
1986.

[93] J.R. Quinlan. *C4.5: Programs for Machine Learning*. Morg\
Kaufmann, 1993.

[94] J.R. Quinlan. Learning efficient classification procedures and their\
plication to chess end games. In Michalski, Carbonell, and Mitch\
editors, *Machine learning: An artificial intelligence approach*. Mor\
Kaufmann, 1993.

[95] M. Saerens, F. Fouss, L. Yen, and P. Dupont. The principal compon\
analysis of a graph and its relationships to spectral clustering. In *EC*\
2004, 2004.

[96] B. Sarwar, G. Karypis, J. Konstan, and J. Riedl. Application of dim\
sionality reduction in recommender systems – A case study. In *A*\
WebKDD Workshop, 2000.

[97] F. Shahnaz, M.W. Berry, V.P. Pauca, and R.J. Plemmons. Docum\
clustering using Nonnegative Matrix Factorization. *Journal on In*\
mation Processing and Management, 42(2):373–386, March 2006.

[98] A. Shashua and T. Hazan. Non-negative tensor factorization with\
plications to statistics and computer vision. In *Proceedings of the 2*\
International Conference on Machine Learning, 2005.

[99] J. Shi and J. Malik. Normalized cuts and image segmentation. *IE*\
Transactions on Pattern Analysis and Machine Intelligence, 22(8):8\
905, 2000.

[100] D.B. Skillicorn. Clusters within clusters: SVD and counterterrorism\
First Workshop on Data Mining for Counter Terrorism and Secur\
2003 SIAM Data Mining Conference, March 2003.

[101] D.B. Skillicorn. Beyond keyword filtering for message and conversat\
detection. In *IEEE International Conference on Intelligence and*\
curity Informatics (ISI2005), pages 231–243. Springer-Verlag Lect\
Notes in Computer Science LNCS 3495, May 2005.

[102] D.B. Skillicorn and C. Robinson. A data-driven protein-structure pre
diction algorithm. Technical Report 2006-509, Queen's University
School of Computing, 2006.

[103] D.B. Skillicorn and N. Vats. The Athens system for novel informa
tion discovery. Technical Report 2004-489, Queen's University Schoc
of Computing Technical Report, October 2004.

[104] D.B. Skillicorn and N. Vats. Novel information discovery for intelligenc
and counterterrorism. *Decision Support Systems*, April 2006.

[105] D.B. Skillicorn and X. Yang. High-performance singular value decom
position. In Grossman, Kamath, Kumar, Kegelmeyer, and Namburu
editors, *Data Mining for Scientific and Engineering Applications*, page
401–424. Kluwer, 2001.

[106] B.W. Smee. A new theory to explain the formation of soil geochemica
responses over deeply covered gold mineralisation in arid environments
Journal of Geochemical Exploration, pages 149–172, 1998.

[107] B.W. Smee. Theory behind the use of soil pH measurements as a
inexpensive guide to buried mineralization, with examples. *Explore*
118:1–19, 2003.

[108] D. Spielman. Spectral graph theory and its applications. Course Notes
www.cs.yale.edu/homes/spielman/eigs/, 2004.

[109] G.W. Stewart. On the early history of the Singular Value Decomposi
tion. Technical Report TR-2855, University of Maryland, Departmer
of Computer Science, March 1992.

[110] P.-N. Tan, M. Steinbach, and V. Kumar. *Introduction to Data Mining*
Pearson Addison-Wesley, 2005.

[111] L.R. Tucker. Some mathematical notes on three-mode factor analysis
Psychometrika, 31:279–311, 1966.

[112] What is Holmes 2. www.holmes2.com/holmes2/whatish2/, 2004.

[113] N. Vats and D.B. Skillicorn. Information discovery within organization
using the Athens system. In *Proceedings of 14th Annual IBM Center
for Advanced Studies Conference (CASCON 2004)*, October 2004.

[114] U. von Luxburg. A tutorial on spectral clustering. Technical Repor
149, Max Plank Institute for Biological Cybernetics, August 2006.

[115] U. von Luxburg, O. Bousquet, and M. Belkin. Limits of spectral clus
tering. In *Advances in Neural Information Processing Systems (NIPS
17*, volume 17, pages 857–864, Cambridge, MA, 2005. MIT Press.

[116] P. Wong, S. Choi, and Y. Niu. A comparison of PCA/ICA for c
      preprocessing in a geoscience application. In *Proceedings of ICA*, pa
      278–283, December 2001.

[117] L. Zelnik-Manor and P. Perona. Self-tuning spectral clustering. In
      vances in Neural Information Processing Systems 16, Cambridge, N
      2004. MIT Press.

[118] M. Zhu and A. Ghodsi. Automatic dimensionality selection from
      scree plot via the use of profile likelihood. *Computational Statistics*
      Data Analysis, 51(2):918–930, 2006.

[119] S. Zyto, A. Grama, and W. Szpankowski. Semi-discrete matrix tra
      forms (SDD) for image and video compression. Technical report,
      partment of Computer Science, Purdue University, 2000.

Index